Springer Aerospace Technology

For further volumes:
http://www.springer.com/series/8613

Tarit Bose

Aerodynamic Noise

An Introduction for Physicists and Engineers

 Springer

Tarit Bose
Ret., Indian Institute of Technology Madras
Sardar Patel Road
Chennai
Tamil Nadu, India

ISSN 1869-1730   ISSN 1869-1749 (electronic)
ISBN 978-1-4939-0196-8   ISBN 978-1-4614-5019-1 (eBook)
DOI 10.1007/978-1-4614-5019-1
Springer New York Heidelberg Dordrecht London

Dedicated to
the original cosmic "Om" sound, which has been replaced over the last hundred years by large sounds from jets and sonic bangs. For the uninitiated, the "Om" is a+a+a+u+m consisting of "Arihanta, Ashiri [that is, siddha], Acharya, Upadhaya, and Munis [sadhus]."

I bow to the Arihants, the perfected human beings.

I bow to the Siddhas, liberated bodiless souls.

I bow to the Acharyas, the masters and heads of congregations.

I bow to the Upadhyayas, the spiritual teachers.

I bow to the spiritual practitioners in the universe, Sadhus.

(read on the Internet)

Preface

My initial contact with the subject of aircraft noise had been rather involuntary. In the autumn of 1968, I was working in the USA in a laboratory financed by NASA, when there was a shift in priorities in the laboratory, and almost overnight I found myself trying to understand the language of a noise engineer. On my return to India in the summer of 1970, one thing led to another, and I was requested to give a series of lectures on the subject. This I did for a number of years to interested students of undergraduate and graduate classes of mechanical and aerospace engineering, for whom a suitable textbook was not available. In fact, an explosive development of a subject like aeronautical engineering associated with a tremendous increase in noise pollution during the twentieth century did not match the development of tools to control the noise. In addition, the subject of the physics of acoustics did not develop till the first half of the nineteenth century and the important work in the second half was the 1885 book *Sound* by Lord Rayleigh. Therefore, I had to prepare my notes from scratch, which were revised again and again and out of which the manuscript of this book developed.

Although a course was offered by me for a number of years, the only option I had was to develop the course notes with little help from the then existing literature. At that time, the Curriculum Development Center of Mechanical Engineering of the Indian Institute of Technology, Madras, offered to give whatever help was needed in developing lecture notes on the subject. This was done with the help of then available production technology of cyclostyling with the mathematical equations and figures written or drawn by hand, giving a very untidy appearance; it was only for limited consumption within the institute. This monograph is an offshoot of the then cyclostyled notes, updated and produced with the help of modern LaTeX technology, although it has been completely rewritten, including all the latest references, and may be considered a new book.

Whereas the subject of aircraft noise is a vast one and is developing at a tremendous pace, I have tried to include as much of recent developments in the subject as possible. The more than a hundred references, most published in the twenty-first century, are testimony to this. This includes the development of new mathematical and computational tools and some experiments, too. For very advanced readers much of the material contained herein may be too elementary, but it is hoped that the monograph will be useful for general readers.

The topics for the monograph were, therefore, selected chiefly to explain the physical processes involved. Sound produced due to the motion of air and its interaction with solid boundaries, which became important mainly after the Second World War, constitutes the main subject of this monograph. In fact, the subject was treated by Lighthill himself for the first time only in 1952. In the introduction to Lighthill's paper, Lighthill points out that the development of the science of acoustics had been rather slow in comparison to other branches of physics and had remained almost static since the 1890s when John Stuart, the third Baron of Rayleigh, wrote his famous treatise on the *Theory of Sounds*. Since then "Rayleigh's world," as described by Lighthill, "of tuning fork, violin, whispering galleries, organ pipes, church bells, singing flames and bird-calls has been stormed, even though by the niceties of acoustic insulation and high-fidelity reproduction, from the air by cacophonous sequence of whines and roars and bangs," which is the main subject of discussion in the paper. The paper focuses on the mechanism of production of noise arising almost entirely from the airflow around a vehicle or engine of an aircraft or a rocket, and hence the name "aerodynamic noise" has come into widespread usage. A related field, aeroelasticity, and a related term, sonic fatigue, have also evolved to characterize the response and breakdown of materials and structures under unsteady aerodynamic and acoustic loadings. As for aerodynamic noise, it is convenient to start by enumerating the breakdown of the major sources of aircraft noise: (1) noise from propellers, nose, and fans; (2) noise from the combustion chamber; (3) noise produced in the jets; (4) noise produced due to interaction with the external surface of the aircraft (boundary layer noise); and (5) the sonic boom. Aviation acoustics is concerned with all these problems of generation of noise at the source, its transmission, and the effect of noise on the receiver, and its understanding will enable design engineers to reduce the adverse effects of noise from aircraft operation. At this stage it may be worthwhile to make some remarks about the definition of noise in general and of a acoustic noise in particular. By definition, noise is any undesirable sound. This is a rather subjective definition and depends on various factors like, for example, previous training, the particular type of sound, physical factors. One also need not bother so much about the about the aircraft noise during the day time, but during night the annoyance due to aircraft noise may be magnified.

Several exercises/problems are given at the end of each chapter. Obviously, these are not the only exercises, and the list can be expanded by the reader.

As was mentioned, the monograph is written mainly for advanced students at the undergraduate or graduate level. It is assumed that the reader already has a good knowledge of gas dynamics and possesses such mathematical tools as complex arithmetic and Fourier analysis.

The manuscript was written and drawings prepared in LATEX by the author. However, writing a book requires the efforts of many people, whose assistance I'd like to acknowledge. My students, who bore the main brunt of listening to the noise I created in my lectures, deserve thanks for their patience and congratulations for surviving. For clarity on the subject, I obtained constant help from Dr. Paul Batten of Metacomp in Agoura Hills, CA, near Los Angeles, and also our former student, Dr. Munipalli Ramakanth of Hypercomp in nearby Westlake Village, CA, for helping me procure a number of relevant papers that were consulted. Dr. Uzun Ali of Florida State University helped me in getting electronic copies of his doctoral thesis, all his papers, and the papers of his doctoral advisors at Purdue University. All American Institute of Aeronautics and Astronautics (AIAA) meeting papers were downloaded, except for a few containing errors, by me directly from AIAA's Web site. My wife Preetishree, daughter Mohua (along with her husband Sumit and their very talkative 5-year old daughter, Mihita), and two sons Mayukh (with his wife, Suzy) and Manjul deserve thanks for maintaining a noise-free environment at home in a world of turbulence and noise.

Currently, the author has retired to Kolkata/India, and most of the book was written there. His e-mail address is bose.tarit@gmail.com, and he would be grateful for any feedback from readers. Thanks are also due to two (anonymous) reviewers for making useful suggestions that greatly improved the book. Springer deserves very hearty thanks for producing this book with their usual loving care.

Chennai, India   Tarit Bose

Contents

Contents xiii

Chapter 1
Introduction

In this chapter, we will discuss those physical aspects of acoustics that are covered in most textbooks on the subject. This includes sound as a wave, waves on a string, aerial waves in tubes and closed rooms (standing waves), periodical phenomena, correlations and spectra, units, and some practical data for aerodynamic noise.

1.1 Sound as a Wave

We write first of the continuity and momentum equations for a nonviscous, 1-dimensional (1D) motion of a gas, in which the square of the velocity terms are neglected, as follows:

Continuity:

$$\frac{\partial \rho}{\partial t} + \frac{\partial (\rho u)}{\partial x} = 0; \tag{1.1}$$

Momentum:

$$\frac{\partial (\rho u)}{\partial t} + \frac{\mathrm{d}p}{\mathrm{d}x} = 0. \tag{1.2}$$

In (1.1) and (1.2), ρ is the (mass) density, p is the pressure, u is the (1D) (gas) velocity, and x and t are the spatial and temporal coordinates, respectively.

Equations (1.1) and (1.2) are differentiated with respect to t and x, respectively, and after subtracting the one from the other, we get

$$\frac{\partial^2 \rho}{\partial t^2} - \frac{\mathrm{d}^2 p}{\mathrm{d}x^2} = 0. \tag{1.3}$$

Let there be a variable c defined by the expression

$$c = \sqrt{\mathrm{d}p/\mathrm{d}\rho} \tag{1.4}$$

T. Bose, *Aerodynamic Noise: An Introduction for Physicists and Engineers*,
Springer Aerospace Technology 7, DOI 10.1007/978-1-4614-5019-1_1,
© Springer Science+Business Media New York 2013

with a unit of speed that we will call the *sonic speed*. Substituting (1.4) into (1.3) we get

$$\frac{1}{c^2}\frac{\partial^2 p}{\partial t^2} - \frac{\partial^2 p}{\partial x^2} = 0. \tag{1.5}$$

Equation (1.5) is known as a second-order hyperbolic partial differential equation with the solution

$$p = p_1(x+ct) + p_2(x-ct). \tag{1.6}$$

Noting from (1.6) that

$$\frac{\partial^2 p}{\partial x^2} = \frac{\partial^2 p_1}{\partial (x+ct)^2} + \frac{\partial^2 p_2}{\partial (x-ct)^2}, \tag{1.7}$$

$$\frac{1}{c^2}\frac{\partial^2 p}{\partial t^2} = \frac{\partial^2 p_1}{\partial (x+ct)^2} + \frac{\partial^2 p_2}{\partial (x-ct)^2}, \tag{1.8}$$

we can show that (1.5) is identically satisfied. It can now be shown that if $(x \pm ct)$ is constant, then the slope of this in the (x,t) plane is

$$-\frac{\mathrm{d}x}{\mathrm{d}t} = \pm c, \tag{1.9}$$

which is an expression for velocity. The first term relates waves moving forward in time but backward in space, whereas the second term relates to waves moving forward in time and forward in space. The two combined directions, $(x \pm ct) =$ constant, are known as characteristics (Fig. 1.1), and the velocity of propagation c for isentropic propagation of information in gas is given by

$$c = \sqrt{\gamma p/\rho} = \sqrt{\gamma RT}, \tag{1.10}$$

where γ is the ratio of specific heat and R is the gas constant. For air at room temperature at sea level (288 K), $c = 340.3\,\mathrm{m/s}$. Thus the speed of sound in gases changes with ambient conditions. In fresh water, at $20\,^\circ\mathrm{C}$, the speed of sound is approximately 1,482 m/s, while the speed of sound in steel at the same temperature is about 5,960 m/s.

The propagation of the disturbance of pressure $p' = p - p_o$ is accompanied by the disturbance of mass density $\rho' = \rho - \rho_o = p'/c_o^2$, where the subscript o refers to the gas at rest. The expression $c = c_o$ indicates the sonic speed for a gas at rest.

Now from the momentum equation (1.2) we write

$$\rho_o\frac{\partial u'}{\partial t} + \frac{\mathrm{d}p'}{\mathrm{d}x} = 0 \tag{1.11}$$

Fig. 1.1 Characteristics in
the (x,t) plane

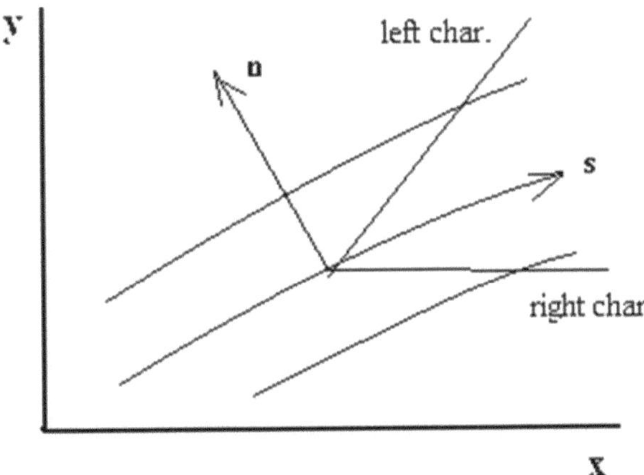

and rewrite the same equation in slightly different form as

$$\frac{\partial u'}{\partial t} = -\frac{c_o^2}{\rho_o}\frac{\partial p'}{\partial x},\qquad(1.12)$$

from which we write further

$$\frac{\partial u'}{\partial t} = -\frac{c_o^2}{\rho_o}\frac{\partial \rho'}{\partial x}\,\mathrm{d}t.\qquad(1.13)$$

Upon integration we now get

$$u' = -\frac{c_o^2}{\rho_o}\int \frac{\partial \rho'}{\partial x}\,\mathrm{d}t.\qquad(1.14)$$

Thus it is shown that in the case of propagation of perturbation in a gas, the pressure perturbation is accompanied by a density perturbation and a small gas "particle" velocity perturbation. The physiologial phenomenon that converts a pressure wave into sound is outside the scope of this monograph, except to imagine the ear as a very sensitive pressure-measuring device (transducer) that can operate over a very wide range of amplitude and frequency of pressure fluctuation. It is enough to point out here that a human body with good hearing capability can hear sound waves with frequencies from a few cycles per second (Hertz) to approximately 10,000 Hz, with the maximum sensitivity at 1,000 Hz. In addition, a person with good hearing can barely hear a sound wave with amplitude $0.0002\,\mathrm{dyn/cm^2} = 2\times10^{-5}\,\mathrm{Nm^{-2}}$. These two bits of information are important for defining the (logarithmic) scales of sound waves.

Although Eqs. (1.5) and (1.6) relate to pressure fluctuations, they can now be written, with the help of (1.4), in terms of the density fluctuation as follows:

$$\rho' = \frac{p'}{c_o^2} = \frac{1}{c_o^2}\left[p_1(x+c_ot)+p_2(x-c_ot)\right]$$

$$= \rho_1(x+c_ot)+\rho_2(x-c_ot).$$

Let us now consider a simple harmonic wave of density in one dimension as

$$\rho' = A_1\cos[\omega(t+x/c_o)]+A_2\cos[\omega(t-x/c_o)], \tag{1.15}$$

where ω is the radian frequency of the traveling wave. Thus the expressions for the perturbation of pressure and particle velocity are given by

$$p' = c_o^2\rho' = c_o^2\left[A_1\cos[\omega(t+x/c_o)]+A_2\cos[\omega(t-x/c_o)]\right], \tag{1.16}$$

$$u' = -\frac{c_o}{\rho_o}\left[A_1\cos[\omega(t+x/c_o)]+A_2\cos[\omega(t-x/c_o)]\right]. \tag{1.17}$$

Of the two parts of the solutions a forward-moving wave will mean $A_1 = 0$ and $A_2 > 0$, and thus ρ', p', and u' are all in phase, whereas a backward-moving wave will mean $A_1 > 0$ and $A_2 = 0$, and for this ρ' and p' are in phase but u' is out of phase. Obviously, except for ρ_o being very small, $|\overline{p'^2}| > |\overline{u'^2}| > |\overline{\rho'^2}|$. For forward-moving waves, if A_2 is the amplitude of the density fluctuation, then $A_2c_o^2$ is the amplitude of the pressure fluctuation and A_2c_o/ρ_o is the amplitude of the velocity fluctuation. For example, the amplitude of pressure for a standard wave is $2\times10^{-5}\,\mathrm{Nm}^{-2}$. With $\rho_o = 1.16\,\mathrm{kg/m^3}$ and $c_o = 330\,\mathrm{m/s}$, the amplitude for the fluctuation of density is $1.83654\times10^{-10}\,\mathrm{kg/m^3}$ and the amplitude for the fluctuation of velocity is $5.22466\times10^{-8}\,\mathrm{m/s}$. Hence the amplitude of the *pressure fluctuation* is much larger than both the *density and velocity fluctuations*.

1.2 The Case of a Stretched String

In the previous section we discussed compressive-expansive type waves. However, for several thousand years, string instruments have been important in producing musical sounds, although the nature of propagation of waves in air is quite different from the propagation of waves on a string, which is accomplished through displacement in the transverse direction along the length of the string. To this end, we will discuss the subject in two subsections dealing with (a) a stretched string subject to continuous stationary transverse load and (b) a string subject to one or more point loads.

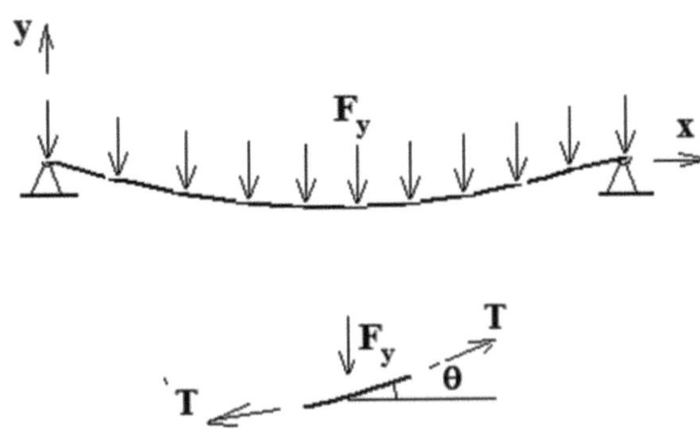

Fig. 1.2 Model of a stretched string element

1.2.1 *Perfect Elastic String Under Tension*

Let us now consider a perfectly flexible string that is stretched with a constant tension T [N] that is so large that this may be much larger than the self-weight of the string distributed uniformly between the two points. Let the string have a mass m per unit length (kilogram per meter) (Fig. 1.2). If Θ is the angle formed with the horizontal axis x, then for small values of Θ one can write

$$\sin\Theta \approx \tan\Theta \approx \frac{dy}{dx}. \tag{1.18}$$

The net force F_y in the y-direction, acting between x and $x + dx$, is

$$F_y dx = T(\sin\Theta_2 - \sin\Theta_1) = \left(T\frac{\partial y}{\partial x}\right)_{x+dx} - \left(T\frac{\partial y}{\partial x}\right)_x$$

$$= \left(T\frac{\partial y}{\partial x}\right)_x + \frac{\partial}{\partial x}\left(T\frac{\partial y}{\partial x}\right)_x + \frac{\partial}{\partial x}\left(T\frac{\partial y}{\partial x}\right) - \left(T\frac{\partial y}{\partial x}\right)_x$$

$$= \frac{\partial}{\partial x}\left(T\frac{\partial y}{\partial x}\right)_x dx$$

$$= T\frac{\partial^2 y}{\partial x^2}dx = mdx\left(\frac{\partial^2 y}{\partial t^2}\right).$$

For constant T, the preceding equation can be written as

$$F_y = \frac{\partial^2 y}{\partial x^2} = \frac{1}{c_o^2}\frac{\partial^2 y}{\partial t^2} = 0, \tag{1.19}$$

where $c_o = \sqrt{T/m}$ (m/s).

If the string is loaded only due to the weight of the string, but otherwise stationary, then $F_y = -mg$, where g is the gravitational acceleration (m/s^2), then we write

$$\frac{\mathrm{d}^2 y}{\mathrm{d}x^2} = -\frac{mg}{T} = -f. \tag{1.20}$$

In the present case, f is a constant (m^{-1}), but it need not be so if the string cross section is variable (m variable). However, a constant f implies that m/T is constant, but if m is also constant, then T is also constant. Integrating the preceding equation from $x = 0$ to $x = L$, where at both ends the string is fixed ($y = 0$), we get, after taking the boundary conditions into account, the relation

$$y = -f\frac{x^2}{2} + C_1 x + C_2 \tag{1.21}$$

with boundary conditions $y = 0$ for both $x = 0$ and $x = L$, so that $C_1 = fL/2$ and $C_2 = 0$. Thus,

$$y = \frac{1}{2} fx(L - x), \tag{1.22}$$

and the string has a parabolic displacement distribution with a maximum displacement of $fL^2/8$ at $x = L/2$.

Now (1.19) is the same as (1.5), and obviously the solution would be the same, that is,

$$y = f_1(x + c_o t) + f_2(x - c_o)t, \tag{1.23}$$

for which we consider the 1D propagation of a vibrating wave with displacement in the lateral direction as

$$\begin{aligned}
y &= A\exp^{\mathrm{j}\omega(t + x/c_o)} \\
&= A\exp^{\mathrm{j}\omega t}\left[\cos\left(\frac{\omega x}{c_o}\right) + j\sin\left(\frac{\omega x}{c_o}\right)\right].
\end{aligned} \tag{1.24}$$

For a string fixed at two ends, we consider only the sinusoidal term and we set the boundary condition $y = 0$ for $x = 0$ and $x = L$ for all values of t, and then the following relations are valid:

$$\frac{\omega L}{c_o} = k\pi, k = 1, 2, 3, \ldots . \tag{1.25}$$

Further, the following relations are valid:

$$\text{Radian frequency:} \quad \omega = \pi c_o k/L \ (\mathrm{rad/s}), \tag{1.26}$$

$$\text{Freq. of oscill.:} \quad \nu = \frac{\omega}{2\pi} = c_o k/(2L) \ (\mathrm{s}^{-1}), \tag{1.27}$$

$$\text{Period of oscill.:} \quad T = \frac{1}{\nu} = \frac{2L}{c_o k} \ (\mathrm{s}), \tag{1.28}$$

$$\text{Wave length:} \quad \lambda = \frac{c_o}{\nu} = \frac{2L}{k} = 2\pi \ (\mathrm{m}). \tag{1.29}$$

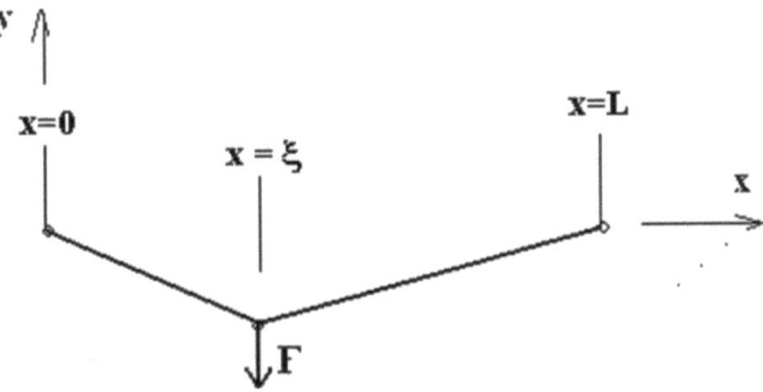

Fig. 1.3 A stretched string under point load

It is understood that k has an *eigenvalue character* and can have only positive integer values, the minimum being one. In the case of $k = 1$, there is no node in any point in the string, and the wavelength is double the length of the string, which is the case of the *fundamental frequency* v_o. Multiples of the fundamental frequency, $2v_o$, $3v_o$, ..., are called *harmonics*, and a slight mistuning of the harmonics produces *beats*.

1.2.2   A Stretched String Under Point Load

A slightly different situation is that of a concentrated stationary transverse load **F**[N] in the y-direction applied to a string of given length L at a given point in the x-direction (Fig. 1.3). As a result, there will be constant tension [N] separately to the left and right of the point of loading T_L and T_R, and there will be different constant slopes in the left and right segments, respectively. In order to use (1.20) for the analysis, however, the right-hand side must be dimensionally compatible with the left-hand side. Therefore, we write $f = FL/(EI)$, where E is the elasticity module (Nm^{-2}) and I is the area moment of inertia (m^4). We obtain the solution on the left- and right-hand sides of the point source position of the string element separately and apply the boundary conditions.

Mathematically, this can be idealized by considering a force limited to an infinitesimally small distance Δ between $x = \xi - \Delta/2$ and $x = \xi + \Delta/2$, as follows:

$$F(x) = \begin{cases} 0, x < \xi - \Delta/2, \\ f/\Delta, \xi - \Delta/2 < x < \xi + \Delta/2, \\ 0, x > \xi + \Delta/2, \end{cases} \tag{1.30}$$

where Δ is a string element of very small length. Distribution of the force F can be replaced symbolically in terms of a so-called delta function, which is given as

$$\delta(x) = \begin{bmatrix} 0, x < \xi - \Delta/2, \\ 1/\Delta, \xi - \Delta/2 < x < \xi + \Delta/2, \\ 0, x > \xi + \Delta/2, \end{bmatrix} \tag{1.31}$$

which, according to Morse and Feshbach [81], is called a *pathological function* since it does not have the properties of continuity and differentiability at $x = \xi$. Around the force point Δ can be made as small as possible ($\Delta \to 0$) with the result that δ is made very large ($\delta \to \infty$). However, it can be integrated and has a fixed value of 1, which is obtained from the integral rule for the delta function

$$\int_{-\infty}^{\infty} F(\xi)\delta(\xi - x)\mathrm{d}\xi = F(\xi) = F. \tag{1.32}$$

Implied in the preceding relation is the validity of a closely related function (*unit step function*) giving the integral properties of the delta function

$$u(x) = \int_{-\infty}^{\infty} \delta(\xi - x)\mathrm{d}\xi = \begin{bmatrix} 0, & x < \xi, \\ 1, & x > \xi. \end{bmatrix} \tag{1.33}$$

From (1.20) we can now write the equivalent differential equation for a concentrated source function as

$$\frac{\mathrm{d}^2 G}{\mathrm{d}x^2} = \delta(\xi - x). \tag{1.34}$$

The solution $G(x)$ satisfies the homogeneous equation $\mathrm{d}^2 G/\mathrm{d}x^2 = 0$ at all points for which $x \neq \xi$, that is, the slope $\mathrm{d}y/\mathrm{d}x$ is constant on either side of the concentration point. The right-hand side is a delta function that, when integrated, gives the unit step function. We can integrate the homogeneous differential equation separately on either side of the source point, matching the deflection at the source point, and the result is given in terms of *Green's function*, $G(x|\xi)$ as follows:

$$y = FG(x|\xi), G(x|\xi) = \begin{bmatrix} x(L - \xi), 0 < x < \xi, \\ \xi(L - x)/L, \xi < x < L. \end{bmatrix} \tag{1.35}$$

We see that the solution is now in a very compact form in terms of Green's function, which has two arguments separated by a vertical line; the first argument gives the point being considered and the second argument gives the position of the point source. For multiple forces, the preceding equation can be generalized as

$$y(x) = \sum F_i G(x|\xi_i). \tag{1.36}$$

1.3   Aerial Waves in Tubes and Closed Rooms

In Sect. 1.1 we wrote down expressions for the fluctuation of density, pressure, and particle velocity in "one dimension." For simple harmonic waves these are rewritten again as follows:

$$\rho' = \frac{1}{c_o^2}\left[A_1 \exp^{j\omega(t+x/c_o)} + A_2 \exp^{j\omega(t-x/c_o)}\right], \tag{1.37}$$

$$p' = c_o^2\rho' = \left[A_1 \exp^{j\omega(t+x/c_o)} + A_2 \exp^{j\omega(t-x/c_o)}\right], \tag{1.38}$$

$$u' = -\frac{1}{\rho_o c_o}\left[A_1 \exp^{j\omega(t+x/c_o)} - A_2 \exp^{j\omega(t-x/c_o)}\right]. \tag{1.39}$$

For standing waves, the solutions of these equations are subject to certain boundary conditions: for an open end exposed to outside atmosphere obviously $p' = 0$, and for a rigid wall $u' = 0$. Therefore, at $x = 0$, if the end is open, then $A = A_1 = -A_2$, and if it is closed, then $A = A_1 = A_2$. Thus if at $x = 0$ the end is open, then

$$\rho' = \frac{2A_1}{c_o^2}\exp^{j\omega t}\sin(kx), \tag{1.40}$$

$$p' = \rho'c_o^2 = 2A_1 \exp^{j\omega t}\sin(kx), \tag{1.41}$$

$$u' = -\frac{2A_1}{\rho_o c_o}\exp^{j\omega t}\cos(kx). \tag{1.42}$$

On the other hand, if the end is closed at $x = 0$, then

$$\rho' = \frac{2A_1}{c_o^2}\exp^{j\omega t}\cos(kx), \tag{1.43}$$

$$p' = \rho'c_o^2 = 2A_1 \exp^{j\omega t}\cos(kx), \tag{1.44}$$

$$u' = -\frac{2A_1}{\rho_o c_o}\exp^{j\omega t}\sin(kx). \tag{1.45}$$

Of course, for the *fundamental frequency*, $k = 1$. If $A_1 = 0$, then

$$p' = A_2 \exp^{j\omega(t-x/c_o)} \simeq A_2 \cos[\omega(t - x/c_o)], \tag{1.46}$$

$$\rho' = \frac{A_2}{c_o^2}\exp^{j\omega(t-x/c_o)} \simeq \frac{A_2}{\rho'c_o^2}\cos[\omega(t - x/c_o)], \tag{1.47}$$

$$u' = \frac{A_2}{\rho_o c_o}\exp^{j\omega(t-x/c_o)} \simeq \frac{A_2}{\rho_o c_o}\cos[\omega(t - x/c_o)]. \tag{1.48}$$

Thus, the amplitude of pressure fluctuation is A_2, the amplitude of density fluctuation is A_2/c_o^2, and the amplitude of velocity fluctuation is $A_2/(\rho_o c_o)$. We consider the velocity fluctuation over one period only; then the *displacement amplitude* is

$$x = \int u'(\mathrm{d}t) = \frac{A_2}{\rho_o c_o 2\pi v} \int \sin(\omega t)\mathrm{d}t = \frac{A_2}{\rho_o c_o 2\pi v}. \tag{1.49}$$

Therefore, the ratio of the displacement amplitude to the pressure amplitude is $1/\rho_o c_o 2\pi v \,(\mathrm{m^3 N^{-1}})$. For the standard pressure wave, the *pressure amplitude* $= 2 \times 10^{-5}\,(\mathrm{Nm^{-2}})$. With $\rho_o \simeq 1.16\,(\mathrm{kg/m^{-3}})$, $c_o = 330\,(\mathrm{m/s})$ in air, and a standard frequency of $1{,}000\,(\mathrm{Hz})$, we get that the displacement amplitude $= 8.3153 \times 10^{-12}\,(\mathrm{m})$, the density amplitude $= 1.83654 \times 10^{-10}\,(\mathrm{kg/m^{-3}})$, and the velocity amplitude $= 5.22466 \times 10^{-8}\,(\mathrm{ms^{-1}})$, and it is the pressure amplitude that dominates.

For the combinations of tube ends open–open, open–closed, closed–open, and closed–closed the final results are as follows:

(a) Open–open or closed–closed:

$$\alpha = \frac{k\pi}{L}, (k = 1,2,3,\ldots); \lambda = \frac{2L}{k}; v = \frac{\omega}{c_o} = \frac{kc_o}{2L}; \tag{1.50}$$

(b) Open–closed or closed–open:

$$\alpha = \frac{(2k+1)\pi}{2L}, (k = 1,2,3,\ldots); \lambda = \frac{4L}{2k+1}; v = \frac{\omega}{c_o} = \frac{(2k+1)c_o}{4L}. \tag{1.51}$$

In Fig. 1.4, the wave forms of pressure and velocity fluctuations for the largest wavelength or smallest frequency (fundamental frequency) for the four cases are shown. It can be seen from these that at a given time, the pressure and velocity fluctuations are $90°$ out of phase of each other. One therefore obtains *nodes* and *loops*. Noting further that the spatial radian frequency $\alpha = \omega/c_o = 2\pi v/c_o = 2\pi/\lambda$, we obtain that the solution gives the relation.

The velocity fluctuation in a node is zero. Thus, if smoke or dust is introduced into a system, then these settle in the nodes (places of vanishing disturbance). However, the pressure fluctuation is at its maximum in nodes.

On the other hand, in a loop the pressure fluctuation is zero (a pressure tap placed there shows no pressure fluctuation), but the velocity fluctuation is maximum.

From (1.40) and (1.42),

$$\frac{\overline{u'^2}}{\overline{\rho'^2}} = \left(\frac{4A_1^2}{\rho_o^2 c_o^2} \frac{c_o^4}{4A_1^2} \right) = \left(\frac{c_o}{\rho_o} \right)^2. \tag{1.52}$$

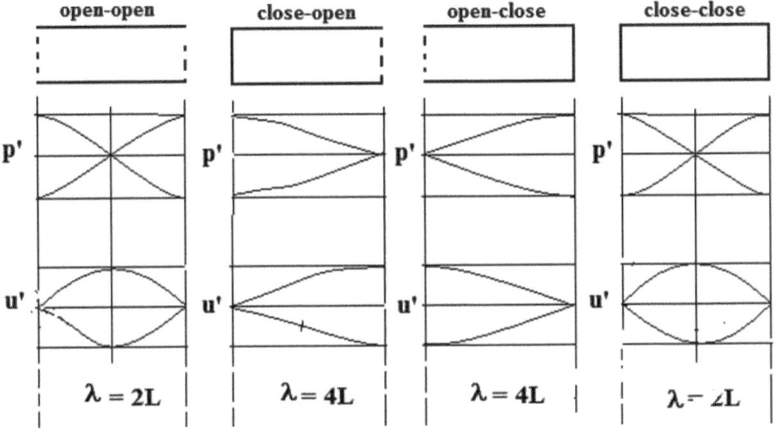

Fig. 1.4 Amplitude of pressure and velocity fluctuations in tubes for largest wavelength

Now the *intensity of sound radiation* is expressed as follows:

$$I\left[\frac{W}{m^2}\right] \sim \rho_o c_o \left(\frac{1}{2}\overline{u'^2}\right) \sim \left(\frac{1}{2}\frac{c_o^3}{\rho_o}\right)\overline{p'^2} = \frac{c_o^3}{\rho_o}\overline{(\rho-\rho_o)^2}. \tag{1.53}$$

Similar results can now be obtained for a closed, rectangular room of length L_x, width L_y, and height L_z. The equations of velocity and pressure fluctuations in such a room are

$$\frac{\partial^2 p'}{\partial t^2} - c_o^2\left[\frac{\partial^2 p'}{\partial x^2} + \frac{\partial^2 p'}{\partial y^2} + \frac{\partial^2 p'}{\partial z^2}\right] = 0, \tag{1.54}$$

$$u_i' = -\frac{1}{\rho_o}\int \frac{\partial p'}{\partial x_i}dt, \tag{1.55}$$

subject to the boundary conditions that the velocity fluctuation is zero at all six (closed) walls.

A general trial simple harmonic solution in three dimensions is

$$p' = \frac{A}{8}\left[\exp^{-jk_x x} + B\exp^{jk_x x}\right]\left[\exp^{-jk_y y} + B\exp^{jk_y y}\right]\left[\exp^{-jk_z z} + B\exp^{jk_z z}\right]. \tag{1.56}$$

Substitution of this equation into (1.54) carries the further requirement that

$$k = \frac{\omega}{c_o} = \left[\sqrt{k_x^2 + k_y^2 + k_z^2}\right]. \tag{1.57}$$

Further, by imposing the condition for velocity fluctuations at the walls, we simplify the equation in the form

$$p' = A\cos(k_x X)\cos(k_y Y)\cos(k_z Z), \tag{1.58}$$

where $X = x/L_x$, $Y = y/L_y$, and $Z = z/L_z$. Further imposing the set of conditions at all boundaries, one gets

$$k_{x,y,z} = \frac{\pi k_{x,y,z}}{L_{x,y,z}}, (k_{x,y,z} = 0,1,2,3,\ldots). \tag{1.59}$$

Thus,

$$k = \frac{\omega}{c_o} = \pi\sqrt{k_x^2 + k_y^2 + k_z^2}, \tag{1.60}$$

$$v = \frac{\omega}{2\pi} = \frac{c_o}{2}k, \tag{1.61}$$

$$\lambda = \frac{2}{k}. \tag{1.62}$$

For standing waves it might appear in a closed room that the acoustic energy associated with pressure fluctuations is at a maximum at some points and zero at other points. Although the acoustic energy is derived in terms of pressure fluctuations only, there are places where the kinetic energy is at a maximum and the acoustic energy is zero. Thus, in a tube or a room, the total energy [consisting of the kinetic and pressure (acoustic) energy] moves back and forth at the speed of sound and is reflected at the closed ends. Every time the sound energy flux is reflected at the wall, there is a loss in the energy due to the transmission and absorption resulting from heat transfer and viscous dissipation in the viscothermal boundary layers at the wall. The heat transfer and viscous dissipation induce a damping that is often stronger than the damping in the bulk of the acoustic field, for which a *coefficient for loss of sound* α is defined.

The viscous dissipation and heat transfer on the surface of particles (for example, water droplets in mist) can induce a strong damping of sound. For pipes with diameter D, which is large compared to the viscous *acoustic boundary layer thickness*

$$\delta_v = \sqrt{2v_f/\omega}.\text{m}, \tag{1.63}$$

where v_f (m^{2}s^{-1}) is the *kinematic viscosity coefficient* of the fluid, the acoustic pressure amplitude of plane waves will decrease over a distance x following the exponential factor $\exp^{-\beta x}$, where the *coefficient for loss of sound*, β, is defined as

$$\beta = \frac{\sqrt{2vv_f}}{Dc_o}\left[1 + \frac{\gamma - 1.}{\sqrt{\text{Pr}}}\right]. \tag{1.64}$$

Here Pr is the Prandtl number and c_o is the sonic speed for the fluid at rest.

Similarly, an unsteady heat transfer at a wall is a source of sound. We consider for simplicity a plane rigid wall and assume that the acoustical velocity is normal to the wall in the y-direction. The characteristic length scale for the *thermal boundary layer* is $\delta_T = \sqrt{2a/\omega}$, where $a = k/(\rho c_p)$ is the *heat diffusivity coefficient* (m^{2}s^{-1}),

k is the *heat conductivity coefficient*, ρ is the *density*, and c_p is the *specific heat at constant pressure*. The *boundary layer thickness* can be defined as the thickness of the region near the wall, in which the unsteadiness of the temperature is due to the heat conduction

$$\rho c_p \frac{\partial T}{\partial t} \simeq k_o. \tag{1.65}$$

Let there be a uniform acoustic energy density E $(\mathrm{Jm^{-3}})$ in a closed room. The intensity of sound on the surface of the wall is $Ec_o/4$ $(\mathrm{Wm^{-2}})$, in which the factor $(1/4)$ consists of products of two factors of $(1/2)$, the one being due to one side of the surface and the other due to integration over all directions (cosine). For a steady-state case inside a room there must be a continuous production of the sound energy q [W], which must be balanced with the energy lost through the total surface area, S, of the room, that is,

$$q = I\beta/S = \beta c_o ES/4. \tag{1.66}$$

On the other hand, the directed sound energy of a point source is $q/(4\pi r^2)$, and the *reflected intensity* is

$$(1-\beta)c_o E - 4(1-\beta)I = 4(1-\beta)q/(\beta S). \tag{1.67}$$

Then the ratio of the reflected sound energy to the directed sound energy is given by the relation

$$\varepsilon = \frac{4(1-\beta)q}{\beta S}\frac{4\pi r^2}{q} = \frac{16\pi r^2}{S}\frac{(1-\beta)}{\beta}. \tag{1.68}$$

In a room of dimension $7.8 \times 4.8 \times 2.4\,\mathrm{m}$ with the radial distance from a single source being $r = 1.5\,\mathrm{m}$, the calculated $\beta = 0.99$. With $\varepsilon = 0.842$ being the typical value of the *loss coefficient* of available material (in terms of decibels), it is equivalent to a reduction of 2.075 only. However, if the surface area is increased manyfold with the help of, for example, a corrugated panel, then it is possible to reduce the value considerably, especially for fiber-glass-panel-lined chambers.

For experimental acoustic measurements in a room, one can have walls with various desirable properties. A heavy material is needed to reduce transmission losses, but for acoustic insulation fibrous materials are needed for good absorption. While for *anechoic chambers* reflection of sound should be reduced as much as possible, a certain amount of reflection is necessary inside the roof of an auditorium so that the sound waves coming from a speaker do not get absorbed into the hair of the audience. Thus the term reverberation time has been introduced; this is the time needed to reduce the sound level by 60 dB in auditoriums with a typical reverberation time of 1 s. A large reverberation time indicates the presence of too much echo, while a small reverberation time indicates too much absorption at the walls and the roof.

1.4 Relations Between Pressure, Density, and Velocity Fluctuations

We showed in the previous section that pressure fluctuation is accompanied by density and particle velocity fluctuations, but the last variable is out of phase with the first two. For a forward-moving simple harmonic wave we can therefore write

$$p' = A_1 \exp^{j\omega(t-x/c_o)}, \tag{1.69}$$

$$\rho' = \frac{p'}{c_o^2} = \frac{A_1}{c_o^2} \exp^{j\omega(t-x/c_o)}, \tag{1.70}$$

$$u' = \frac{jp'}{\rho_o c_o} = \frac{jA_1}{\rho_o c_o} \exp^{j\omega(t-x/c_o)}, j = \sqrt{-1}, \tag{1.71}$$

where A_1 is the amplitude of pressure (Nm^{-2}). For an acoustic wave moving in air at room pressure and temperature, the density of the air at rest can be considered approximately equal to one, and thus it can be seen that the amplitude of the pressure fluctuation is the largest and the amplitude of the density fluctuation is the smallest, which we also showed earlier. We can also evaluate the root mean square (designated by "$<>$") of the three variables and obtain the following expressions:

$$<p'> = \frac{A_1}{2}, \tag{1.72}$$

$$<\rho'> = \frac{<p'>}{c_o^2} = \frac{A_1}{2c_o^2}, \tag{1.73}$$

$$<u'> = \frac{<p'>}{\rho_o c_o} = \frac{A_1}{\rho_o c_o}. \tag{1.74}$$

1.5 Periodic Phenomena

In the previous two sections we discussed cases of simple harmonics, although it is known that sound waves from most sources, like most periodic phenomena in nature, are far too complex to be described with simple harmonics. It is, therefore, assumed that vibrations take place simultaneously at a number of different frequencies or radian frequencies, which are the integer multiple of the fundamental radian frequency, ω. When radian frequencies are excited simultaneously, the resulting vibration has a complex wave form, while the period, T, remains the same as that of the fundamental frequency given by $T = 2\pi/\omega$. We discuss the phenomena here very briefly, but for more detailed discussion any good book on mathematics may be consulted.

Now let there be a periodic phenomenon given by a series expression (textit-Fourier series)

$$F(t) = a_0 + a_1 \exp^{j\omega t} + a_2 \exp^{2j\omega t} + \cdots + a_n \exp^{nj\omega t} + \cdots$$
$$+ a_{-1} \exp^{-j\omega t} + \cdots + a_{-n} \exp^{-nj\omega t} + \cdots$$
$$= \sum_{n=-\infty}^{n=\infty} a_n \exp^{nj\omega t}. \tag{1.75}$$

The *coefficients* (known as *Fourier coefficients*) are given by the relation

$$a_n = \frac{1}{T} \int_{-T/2}^{T/2} F(t) \exp^{-jn\omega t} dt. \tag{1.76}$$

If the values of a_n are plotted against $n\omega$, that is $\Delta\omega = 2\pi/T$ apart, then it can be shown that the coefficients converge rapidly as n is increased in either direction. Thus, as $T \to \infty, \Delta\omega \to 0$, the Fourier series given by (1.75) is converted into a *Fourier integral*.

Let us now introduce a *dummy variable* δ into (1.76) to replace t but having the same character as time t. Therefore,

$$F(t) = \sum_{n=-\infty}^{n=\infty} \left[\frac{1}{T} \int_{-T/2}^{T/2} F(\delta) \exp^{-jn\omega\delta} d\delta \right] \exp^{2\pi jn\omega\delta}$$
$$= \sum_{n=-\infty}^{n=\infty} \left[\Delta v \int_{-T/2}^{T/2} F(\delta) \exp^{-2\pi jnv\delta} d\delta \right] \exp^{2\pi jnvt}. \tag{1.77}$$

For $T \to \infty$, (1.77), which is the *Fourier integral*, becomes

$$F(t) = \int_{-\infty}^{\infty} \exp^{2\pi jvt} dv \int_{-\infty}^{\infty} F(\delta) \exp^{-2\pi jv\delta} d\delta. \tag{1.78}$$

The expression on the right-hand side of the second integral sign is defined as $G(v)$. Further, the dummy variable is replaced back again. One gets, therefore, a set of two equations, so-called *Fourier transforms*, as follows:

$$F(t) = \int_{-\infty}^{\infty} G(v) \exp^{2\pi jvt} dv, \tag{1.79}$$

$$G(v) = \int_{-\infty}^{\infty} F(-t) \exp^{-2\pi jvt} dt. \tag{1.80}$$

In (1.79) and (1.80), if integration is performed from 0 to ∞, then

$$F(t) = \int_{0}^{\infty} G(v) \exp^{2\pi jvt} dv, \tag{1.81}$$

$$G(v) = \int_{0}^{\infty} F(-t) \exp^{-2\pi jvt} dt. \tag{1.82}$$

Let

$$F(t) = \exp^{-t/\lambda_t};$$ (1.83)

then

$$G(v) = -\frac{1}{\left(\frac{1}{\lambda_t} + 2\pi j v\right)}.$$ (1.84)

Equations (1.79) and (1.80) are used to determine the spectral characteristic of the radiated sound.

1.6 Probability, Correlations, and Spectra

Let there be fluctuations of a certain property, q, which may be either density or pressure or velocity. It was shown in Sect. 1.3 that density and pressure fluctuations are related by the relation $\rho' = c_o^2 p'$, whereas these are related to velocity fluctuations in a somewhat complicated manner because the latter can be $90°$ out of phase in space with respect to the former. In a study by Parthasarathy et al. [89], simultaneous measurements were made of density fluctuations with *crossed laser beams*, pressure fluctuations with very sensitive *microphones* as well as static probes, and velocity fluctuations with *hot-wire anemometry*. The results showed large discrepancies in the measurements of normalized correlations. However, in principle, measurements of any of the foregoing properties, $q = q(x,t)$, can be made.

A record of such measurements can be made either at one particular point x as a function of time t or by simultaneous measurements at two different nearby points. The results will be similar in both cases, and therefore we write a general expression $q = q(\varphi)$, where φ is either a space or time variable and the other is kept fixed. The general property variable q may consist of an average value $\bar{q}$ independent of φ and a fluctuating component with respect to the average value, $q'(\varphi)$ (Fig. 1.5).

Therefore, we write

$$q = \bar{q} + q'(\varphi).$$ (1.85)

The average $\bar{q}$ is defined by the relation

$$\bar{q} = \lim_{\Omega \to 0} \left[\frac{1}{\Omega} \int_{\Omega=0}^{\Omega} q \, d\Omega \right],$$ (1.86)

where Ω is the volume element.

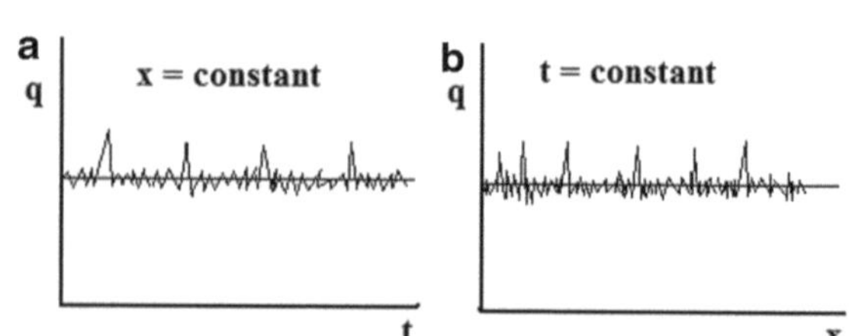

Fig. 1.5 Sample records of a property q (shown schematically) as a function of (**a**) time for x = const and (**b**) x for t = const

It is obvious from (1.85) and (1.86) that

$$\bar{q}' = \lim_{\Omega \to 0} \int_{\Omega=0}^{\Omega} (q - \bar{q}) \mathrm{d}\Omega = \bar{q} - \bar{q} = 0. \tag{1.87}$$

If, however, the perturbed quantity becomes negative, then there can be difficulties in evaluating (1.87). In such cases it is better to take the mean of a root square value (*root mean square*); the square of the root mean square is defined as

$$(\tilde{q}) = \sqrt{\overline{q'^2}} = \lim_{\Omega \to 0} \left[\frac{1}{\Omega} \int_{\Omega=0}^{\Omega} (q')^2 \mathrm{d}\Omega \right] = \overline{(q - \bar{q})^2}. \tag{1.88}$$

While the preceding integral requires continuous signals, one may conduct experiments in which there may be signals at discrete times. For example, let there be a total of N discrete observations. This can happen by observing either at one point for a total N times at equal intervals or simultaneously by introducing N probes at equal distance. Out of these N values, we may group the observed values within a specified interval, Δq, which we may observe n_k times:

$$F(t) = \int_{-\infty}^{\infty} G(v) \exp^{2\pi \mathrm{j} v t} \, \mathrm{d}v, \tag{1.89}$$

$$G(v) = \int_{-\infty}^{\infty} F(-t) \exp^{-2\pi \mathrm{j} v t} \, \mathrm{d}t. \tag{1.90}$$

Actual numerical evaluation is done with the help of the *fast Fourier transform (FFT)*, a description of which is beyond the scope of this book.

Let the range of values be subdivided into q_k divisions of interval Δq groupings. Hence we may write

$$\sum_{k=1}^{r} n_k = N, \tag{1.91}$$

and the probability of having at least a value of q_k is

$$W_k = \frac{1}{N} \sum_{k}^{r} n_k, \tag{1.92}$$

which means, obviously, that $W_1 = 1$. It is evident that the probability, W_k, depends on q_k and Δq, and $W_k \to 0$ as $\Delta q \to 0$. However, their ratio reaches a limiting value, which is the *probability density function*

$$f(q) = \lim_{\Delta q \to 0} \left| \frac{\Delta W_k}{\Delta q} \right| = \frac{\mathrm{d}W}{\mathrm{d}q}, \tag{1.93}$$

and upon integration the probability is

$$W = \int_{0}^{q} f(q) \mathrm{d}q. \tag{1.94}$$

Of course, it is evident that

$$\int_0^\infty f(q)\mathrm{d}q = 1, \tag{1.95}$$

and an average can be computed from the relation

$$\bar{q} = \int_0^\infty qf(q)\mathrm{d}q = \int_0^1 q\mathrm{dW}. \tag{1.96}$$

While the previous discussions pertain to the probability of events within certain (time or space) intervals, much use has been made of correlations, following an original idea of Taylor [112], of fluctuation of properties at neighboring points x and $x+\xi$. The idea leads to the definition of a *correlation volume*. In turbulent flows, for example, Taylor showed that the diffusion of a (fluid) particle starting from a point depends on the *correlation* between the velocity of the particle at any instant t and that of the particle after a time $t + \tau$.

Thus the general relation for a correlation is

$$R(\xi,\tau) = \overline{q'(x,t)q'(x+\xi,t+\tau)}, \tag{1.97}$$

where the "bar" denotes the time average and ξ and τ are space (distance) and time intervals, respectively.

Equation (1.97) has two variables and is very difficult to operate. Thus either ξ or τ is set equal to zero. We introduce, therefore, a variable, φ, that can be either a space or time coordinate, and we write

$$R(\varphi') = \overline{q'(\varphi)q'(\varphi+\varphi')}. \tag{1.98}$$

Let the fluctuations be given in terms of the Fourier series

$$q'(\varphi) = \sum_{k=-\infty}^{\infty} A_k \exp^{jk\omega_o \varphi}, \tag{1.99}$$

$$q'(\varphi+\varphi') = \sum_{m=-\infty}^{\infty} A_m \exp^{-jm\omega_o(\varphi+\varphi')}. \tag{1.100}$$

Therefore, for the correlation we write

$$R(\varphi') = \overline{q'(\varphi)q'(\varphi+\varphi')} = \sum_{k,m} \overline{A_k A_m} \exp^{j[(k-m)\omega_o\varphi - m\omega_o\varphi']}, \tag{1.101}$$

which is dependent on φ' only if $k = m$ and $\sum_{k,m,k \neq m} \overline{A_k A_m} = 0$. Thus,

$$R(\varphi') = \sum_k \overline{A_k A_k} \exp^{-jk\omega_o\varphi'}, \tag{1.102}$$

which can be normalized by dividing by $R(0)$ (Fig. 1.6a, b).

Fig. 1.6 Correlation

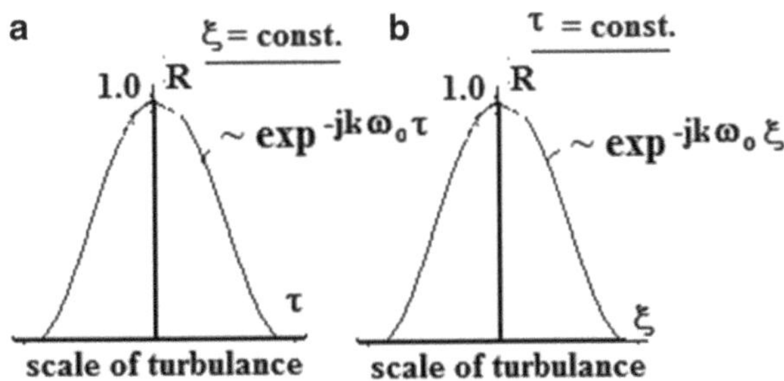

The shape of the correlation coefficient against the distance between the two positions but measured simultaneously yields useful information regarding the scale of turbulence and, thus, information regarding the turbulent noise-producing zone. On the other hand, if ξ is set equal to zero, then one gets the *autocorrelation*, which is, in fact, a Fourier series transform of the power spectral density of the signal and may be used to determine the spectrum.

Let there be a correlation of density fluctuations at two different times with time delay τ, which is

$$B(\tau) = \overline{\rho'(x,t)\rho'(x,t+\tau)}, \mathrm{kg^2 m^{-6}}. \tag{1.103}$$

Now with (1.80) we can write the spectral relation

$$G(v) = \int_{-\infty}^{\infty} B(\tau)\exp^{-2\pi jv\tau} d\tau \tag{1.104}$$

to determine the spectrum. It can be shown that in a homogeneous fluctuating field the envelope of all cross-correlation curves – each of them drawn for a given *space gap* ξ as a function of the *time delay* τ – is in fact the *autocorrelation* of the signal that would be seen by an observer traveling with the turbulence, so that its *Fourier cosine transform* is the *power spectral density* of the fluctuating quantity relative to areas moving with the stream.

Thus, one can write the two expressions for the *intensity of sound* and the power spectral intensity of sound as

$$I\left[\frac{W}{m^2}\right] \simeq \frac{c_o^3}{\rho_o}\overline{\rho'^2} \simeq \frac{c_o^3}{\rho_o}B(0), \tag{1.105}$$

$$\frac{dI}{dv}\left[\frac{J}{m^2}\right] \simeq \int_{-\infty}^{\infty} B(\tau)\exp^{-2j\pi v\tau} d\tau. \tag{1.106}$$

1.7   Units

It may be pointed out that the fluid particle displacement Δ involved in the *acoustic wave motion* can become much smaller than the *molecular mean free path* λ_m. For example, for harmonic waves

$$\Delta = \frac{p'}{2\pi v \rho_o c_o},\tag{1.107}$$

which for $p' = 5 \times 10^{-5}$ Pa and $v = 1{,}000$ Hz gives a value of $\Delta = 10^{-11}$ m. On the other hand, from the kinetic theory of gas, for air under standard ambient conditions, $\lambda_m \simeq 10^{-5}$ (m).

We now take a heuristic approach to calculate the *intensity of sound, I,* as follows:

$$I[\mathrm{Wm^{-2}}] = \rho_o c_o \left[\frac{1}{2} <u'>^2\right] = \frac{c_o^3 <\rho'>^2}{\rho_o} = \frac{<p'>^2}{\rho_o c_o}.\tag{1.108}$$

In the preceding equation we made use of (1.72) and (1.73), and "$<>$" means the root mean square value.

As discussed in (1.103), we can define for density fluctuations an autocorrelation with time delay τ, and we can replace the square of the density fluctuation term in (1.108) by the correlation to get

$$I[\mathrm{Wm^{-2}}] = \frac{c_o^3 <\rho'>^2}{\rho_o} = \frac{c_o^3}{\rho_o} B(0).\tag{1.109}$$

From (1.104) we can thus write the *spectral distribution of intensity* as

$$I_v = \frac{\mathrm{d}I}{\mathrm{d}v} = \int_{-\infty}^{\infty} B(\tau) \exp^{-2\pi j v \tau} \mathrm{d}\tau.\tag{1.110}$$

The measurement of the intensity of sound is made, in general, with respect to the *sound pressure level* (SPL) of a standard source with a *standard pressure amplitude* of fluctuation $0.0002\,\mathrm{dyn.cm^{-2}}$, that is, 2×10^{-10} bar, and a discrete *standard frequency* of 1,000 cycles per second (Hz). The reason for standardizing 1,000 Hz as the frequency of a standard measurement is that the human ear is most sensitive around this frequency, although the frequency range of hearing is between a few Hertz and 20,000 Hz. Further, at 1,000 Hz frequency the barely audible amplitude of the SPL for a young person with good hearing is $0.0002\,\mathrm{dyn.cm^{-2}}$. Thus for a standard source (designated by an asterisk) the pressure difference with the atmosphere is

$$p^* - p_o = 2 \times 10^{-5} \cos(2{,}000\pi t).\,\mathrm{Nm^{-2}},\tag{1.111}$$

with the root mean square of the fluctuating standard pressure being 10^{-5} ($\mathrm{Nm^{-2}}$). Taking values under room conditions as $\rho_o = 1.2$ ($\mathrm{kg.m^{-3}}$) and $c_o = 330$ ($\mathrm{m.s^{-1}}$), the intensity of standard sound becomes

$$I^* = \frac{10^{-10}}{330.0 \times 1.2} = 2.5 \times 10^{-12} \approx 10^{-12},\mathrm{W.m^{-2}}.\tag{1.112}$$

The *acoustic power* is obtained by integrating the intensity of sound over the surface of a sphere. Taking the radius of the sphere as 1 m, the acoustic power of a standard source is

$$P^* = 4\pi I^* = 10^{-11}\pi \approx 10^{-11}, \text{W}. \tag{1.113}$$

The intensity of sound in *decibel* (dB) is, by definition,

$$I[\text{dB}] = 10\log_{10}\left[\frac{<p'>^2}{<p^*>^2}\right] = 10\log_{10}\left[\frac{c_o^3 <\rho - \rho_o>^2}{10^{-12}\rho}\right]. \tag{1.114}$$

A corresponding formula for the acoustic power in decibels is

$$P[\text{dB}] = 10\log_{10}\left(\frac{P[\text{W}]}{10^{-11}[\text{W}]}\right). \tag{1.115}$$

These are explained in what follows.

Instead of writing the SPL in terms of the pressure fluctuation, one could write it also in terms of energy fluctuation in the following manner. It is known that if there is a pressure change, the equivalent energy change (*enthalpy difference*) to compress the unit mass of gas (air) is

$$\Delta h = \frac{1}{\rho_o}\Delta p \approx \frac{p - p_o}{\rho_o}. \tag{1.116}$$

Thus, we can also write the *intensity of sound* as

$$I = 10\log_{10}\frac{\overline{(p - p_o)^2}}{\overline{(p^* - p_o)^2}} = 10\log_{10}\frac{\overline{(\Delta h)^2}}{\overline{(\Delta h^*)^2}}, \tag{1.117}$$

with

$$\frac{\overline{(p^* - p_o)^2}}{(\rho_o c_o)} = \frac{A_p^2}{2\rho_o c_o} = \frac{4\times 10^{-10}}{2\times 330 \times 1.2}\frac{(\text{N/m})^2}{(\text{m/s}).(\text{kg/m}^3)}$$
$$= 0.505 \times 10^{-12} \approx 10^{-12}[\text{Wm}^{-2}]. \tag{1.118}$$

Thus,

$$I = 10\log_{10}\left[\frac{c_o^3\overline{(\rho - \rho_o)^2}}{\rho_o \times 10^{-12}}\right], \text{dB}. \tag{1.119}$$

Further, for $r = 1$ m,

$$P = I \times 4\pi r^2 = 0.505 \times 10^{-2}[\text{W/m}^2] \times 4\pi[\text{m}^2] = 0.63 \times 10^{-11}[\text{W}] \approx 10^{-11}[\text{W}]. \tag{1.120}$$

Since evaluation of the *intensity of sound* (in decibels], even for *broadband noise*, requires a comparison of the fluctuations at any frequency with the standard fluctuation at the standard frequency, there are difficulties in evaluation since, as was already mentioned, the human ear is not equally sensitive at all frequencies (or wavelengths) and the ears of different people may not be equally sensitive. Thus was developed a new unit, the *perceived noise decibel* (PNdB), which is identical to the *A-weighting method*, defined such that the PNdB rating of a complex sound should approximate the decibel rating of a 1,000-Hz octave band. The scale requires that the SPL be measured in each of nine contiguous frequency ranges. It may be mentioned also that for pure tones, the unit of loudness in comparison to a discrete frequency sound at 1,000 Hz is the *phon*, which is similar to the PNdB rating of a complex sound.

On the other hand, the *effective perceived noise level* (EPNL) accounts for the duration and presence of discrete frequency tones. It involves a correction factor that adds to the PNL when the noise spectrum contains discrete tones. It also includes a correction obtained by integrating the PNL over a 10-s time interval.

The perception of sound in any organism is limited to a certain range of frequencies. For humans, hearing is normally limited to frequencies between 20 and 20,000 Hz (20 kHz), although these limits are not definite. The upper limit generally decreases with age, and sound reception may be different if the waves hit the eardrum directly or through the bones behind the ear. Other species have different ranges of hearing. For example, dogs can hear vibrations higher than 20 kHz but are deaf to anything below 40 Hz.

A PNdB is obtained by comparing sound received through *headphones* by a sound level of a pure tone at a given frequency in one ear with the other ear muffled by a specific sample of sound. Hence the specific frequency distribution of the sample sound level is important. Conversion charts that have been prepared in various countries differ slightly from each other, and the values in these charts depend on the noise source, for example, whether it is from a jet engine, stationary gas turbine, piston engine, or some other source.

Charts for conversion between phons and decibels are available as an average subjective measurement of a large number of test subjects at different frequencies of sound (Fig. 1.7). It has been found that the ear cannot respond to sound waves at frequencies above 20,000 Hz. Figure 1.8 gives some examples of levels of sound in a modern environment and equivalent amplitude of pressure fluctuation of a standard source. The ear cannot stand continuous exposure to a sound level larger than 100 PNdB for an indefinite period without serious hearing impairment. Further, the amplitude of pressure fluctuations with respect to the ambient pressure is so small that at all sound pressure levels of interest it is possible to assume that the process is isentropic and the sound waves propagate at the local isentropic sonic speed.

The salient feature of a high-pitch (high-frequency) noise is that it is relatively noisier at the same decibel level as a low-pitch noise. Thus a turbojet, with its strong high-frequency content, is rated for cases where it has an overall sound pressure level of 100 dB to be about 6 dB noisier than a propeller. For typical equipment, therefore, special charts for conversion between decibels and perceived noise decibels are needed that could be used for a particular piece of equipment only.

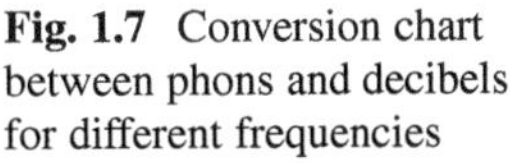

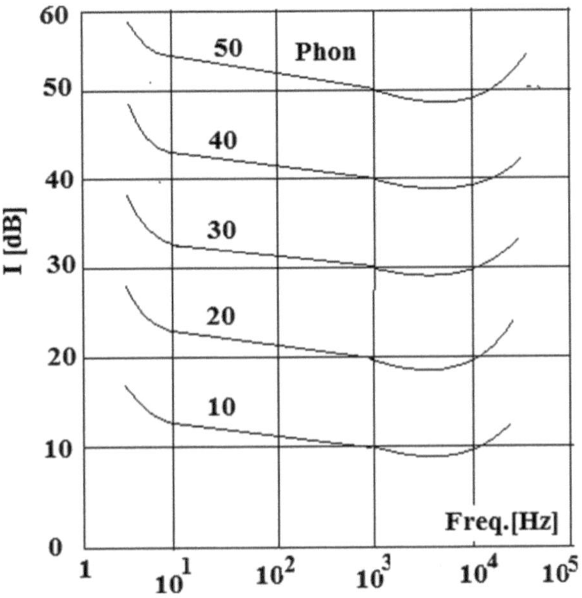

Fig. 1.7 Conversion chart between phons and decibels for different frequencies

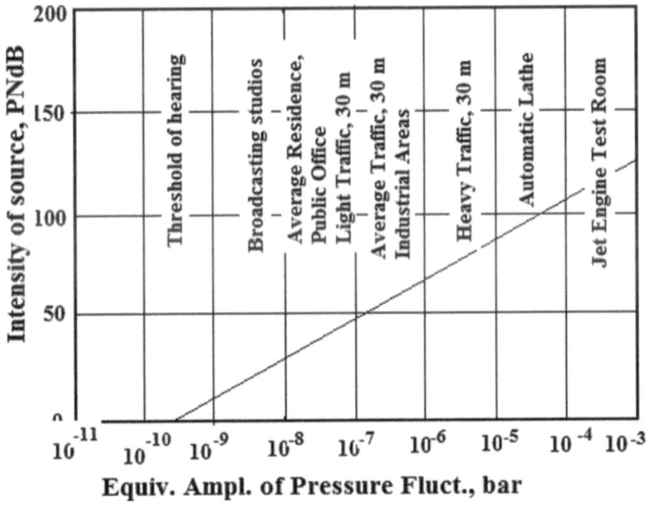

Fig. 1.8 Examples of sound level

The frequency of takeoffs and landings as well as their individual noisiness in perceived noise decibels is a strong factor in public acceptability. Figure 1.9 is, for example, the typical loudness of an aircraft noise with a straight engine (Lloyd [64]) at the London Airport. It shows that during takeoff the sound pressure level is approximately 8–10 dB greater than during landing.

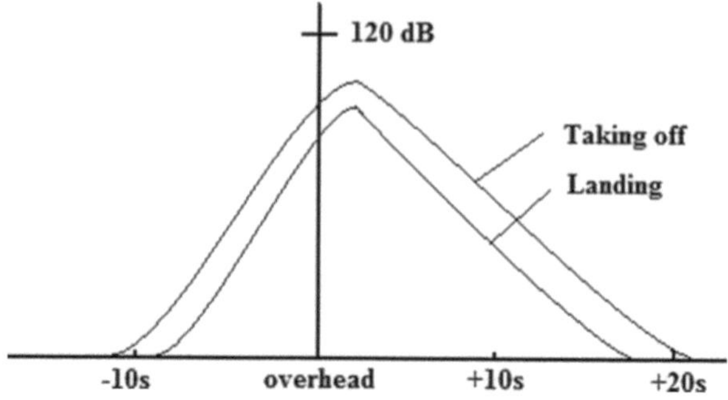

Fig. 1.9 Measurement of a typical aircraft noise (Boeing 707, 150 m overhead [64]

While the preceding figure gives the overall distribution of the SPL, a typical spectral distribution of the SPL of a fanjet engine for subsonic flows through blades is given in a paper by Morley [81], in which some of the peaks within the broadband jet noise spectrum are given; the peak noise represents simple multiples of the product of the number of fans and the rotation of the shaft per unit time due to crossing of the blades in a row in the wake of the blades of the previous row. On the other hand, the SPL of a fanjet engine for a subsonic flow inside the engine is given. Thus the latter can be thought of as emitting from fan stages, and an analysis of the spectrum is, therefore, a very convenient way to detect the faults in a particular fan row. Since the ear can respond to sound waves below 20,000 Hz only, some attempts have been made to shift the frequency, for example, by changing the number of blades or by having a curved leading edge.

The noise index, developed for aircraft noise to test public acceptability, is given by the relation

$$NO = \text{Average PNdB} + \log_{10}(NO) - 80, \qquad (1.121)$$

where NO is the number of occurrences per day. The idea is that if there are a few occurrences per day, one could probably tolerate the situation better than a less noisy sound that occurs every few minutes. However, the preceding expression does not include the annoyance level in the public's mind when people are sleeping at night, although many of the world's busy airports do not allow takeoff or landing at night.

For a very busy airport in the West, there may be, during peak hours, one aircraft landing or taking off every minute. This makes NO = 720 flights over a 12-h period and very few flights at night. Taking an average of 120 PNdB, this means that at such a busy airport, $NI = 120 + 15\log_{10}(720) - 80 = 82.8$. So the maximum tolerable noise index, far from residential homes, might be around 80.

The U.S. Environmental Protection Agency (EPA) uses a Day-Night Average A-Weighted Sound Level metric (DNL) as a method for predicting the effect on a population of long-term exposure to environmental noise.

1.8 Examples of Aircraft Noise

In the previous section we discussed jet noise on the one hand and, on the other, noise from the compressor or fan in a turbojet. The latter can compete with the former, especially at reduced revolutions per minute of the turbomachinery. When an aircraft comes in for a landing, an observer below the flight path hears two maxima – one slightly before the aircraft is overhead due to the fan noise from the engine inlet, the other emanating from the jet after the aircraft has passed.

Although the axial-flow fan used in an aircraft bears a familial resemblance to a propeller, the propeller noise theory applied to a fan can greatly overestimate the noise because of the interaction with and radiation from the duct inlet. It has been found that pressure noise distribution is less effective can decay exponentially in their passage through the duct, whereas in a supersonically spinning mode the noise distribution is less effective upstream and is zero under *choking* conditions.

For certain propulsion units like those of piston engines and pulse jets, combustion noise is important and may be due to one of three mechanisms. Firstly, the turbulence interaction with the reaction of combustible gas may be referred to as *direct combustion noise*. Secondly, the combustion process causes a change in the noise pattern, which may be called *indirect combustion noise*. A third tentative mechanism is the creation of noise by the convection of hot spots generated by the combustion process, which may be called the *entropy noise*. It is known that below a certain jet velocity, say 300 m/s in turbopropulsion systems, this noise due to combustion may tend to become more important, behaving basically as *monopole* sources.

For a supersonic flying aircraft, shocks generated at the bow and tail of the aircraft may generate a big *sonic boom*. However, there exists a cutoff Mach number below which the shock pattern may not reach the ground due to atmospheric refraction when the ground speed of the shock pattern, corrected for the wind gradient, just matches the speed of sound on the ground.

Jet noise and boundary layer noise are due to the random pressure fluctuations of the flow because of turbulence. For subsonic jets, discussed later, Lighthill's formula can provide a good semiempirical relationship for calculating the acoustic power, in which the proportionality factor is substituted from experimental results and is given by the following equation':

$$P[W] = 10^{-4} \rho_o U_{\text{jet}}^8 D_{\text{jet}}^2 / c_o. \qquad (1.122)$$

1.9 Examples of Vehicular Horns and Whistles

Oliver Lucas of Birmingham, UK, was apparently the first person to develop, in 1910, a standard electric car horn, driven by a flat circular steel diaphragm with an electromagnet acting on it and attached to a contactor that repeatedly interrupted the electromagnet sound levels at approximately 107–109 dB at around 400 Hz drawing 5–6 A electric current.

Truck (or lorry) horns are often not activated electrically like in a car but are purely acoustic, activated by air from an air compressor that many trucks already have on board to operate the airbrakes. Such horns have lower frequencies at 125–180 Hz, but sound levels are considerably higher at 117–118 dB.

Train horns may have a chord with notes sounded together and are operated by compressed air from the air system; in a steam engine, steam is released through a nozzle. Train whistles generally produce three or four different frequencies and different codes, for example, to come to a stop, to start, or to pass through an unguarded crossing with a road. Interested readers may wish to listen to the sounds of different train whistles at the Web site of the *Museum of the American Railroad Train Whistles*.

Ships signal to each other and to shore using horns. The modern *International Maritime Organization* specifies that ship horn frequencies must be within a range of 70–200 Hz for vessels over 200 m in length; traditionally, the lower the frequency, the larger the ship.

1.10 Sounds of Music

The world is full of music. There are many different kinds of musical instrument, including *string instruments* like the *guitar*, *violin*, *sitar*, and *sarod*; *percussion instruments* such as *drums* and (Indian) *tablas*; and wind instruments like *whistles* and *flutes* (Fig. 1.10); the latter can be made of any material like metal, wood, or bamboo. All these instruments must be played within definite frequencies and wavelengths.

Musical scales are a sequence of musical notes in ascending and descending order. Most scales are *octave*, repeating their patterns of notes at every octave. The commonly used scales are separated by whole- and half-step intervals of tones and semitones. If the beginning of each octave frequency is doubled for each subsequent octave, it is obvious that beginning of each octave wave length is halved for each

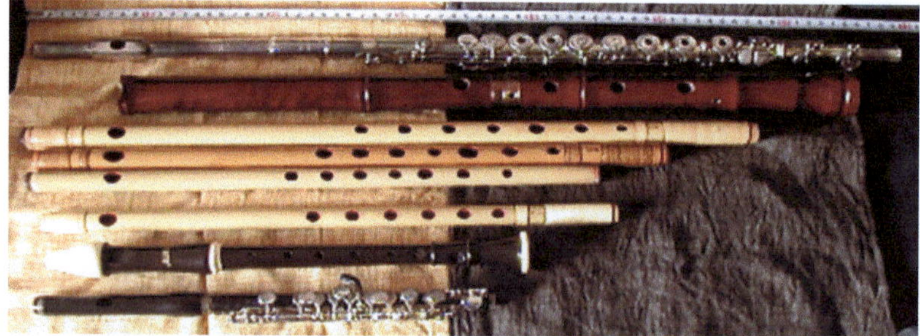

Fig. 1.10 Examples of various flutes in the world (*taken from http://commons.wikimedia.org/ wiki/File:Shinobue_and_other_flutes.jpg*) [64]

subsequent octave, since wave frequency multiplied with wave length is equal to the sonic speed. In many scales, an octave has a major division of seven notes; hence, both end notes have the same octave name, although there can be a larger number of notes in an octave; for example, in the classical Arabic scale there may be 17, 19, or even 24 notes, or the *chromatic scale* has 12 notes, which are called traditionally as within an octave.

In the seven-note system, A to G, between 440 and 880 Hz, the frequency of the individual notes are 440.0, 493.88, 523.25, 587.33, 659.26, 701.0, and 783.99 Hz, and the next octave starts at 880.0 Hz. These frequencies can be converted into wavelengths if one knows the sonic speed. For example, for a sonic speed of $c = 330.0 \text{ ms}^{-1}$, the wavelengths (m) are 0.75, 0.668, 0.631, 0.562, 0.500, 0.471, 0.421, and 0.375. Obviously, neither the frequency nor the wavelength distribution is linear.

As was reported in an article on the Internet titled "A Brief History of Tuning," reprinted from *Fidelio Magazine* and disseminated by the *Schiller Institute* in Washington, DC, the first frequency in an octave was not always 440 Hz:

"The first explicit reference to the tuning of middle C at 256 oscillations per second was probably made by a contemporary of J.S. Bach. It was at that time that precise technical methods developed making it possible to determine the exact pitch of a given note in cycles per second. The first person said to have accomplished this was Joseph Sauveur (1653–1716), called the father of musical acoustics. He measured the pitches of organ pipes and vibrating strings, and defined the "ut" (nowadays known as "do") of the musical scale at 256 cycles per second.

J.S. Bach, as is well known, was an expert in organ construction and master of acoustics; in addition he was in constant contact with instrument makers, scientists, and musicians all over Europe. Thus, we can safely assume that he was familiar with Sauveur's work. In Beethoven's time, the leading acoustician was Ernst Chladni (1756–1827), whose textbook on the theory of music explicitly defined C = 256 as scientific tuning. Up through the middle of the last century, C = 256 was widely recognized as the standard *scientific pitch* or *physical pitch*.

In fact, A = 440 has never been the international standard pitch, and the first international conference to impose A = 440, which failed, was organized by Nazi Propaganda Minister Joseph Goebbels in 1939. Throughout the seventeenth, eighteenth, and nineteenth centuries, and in fact into the 1940s, all standard U.S. and European textbooks on physics, sound, and music took as given the *physical pitch* or *scientific pitch* of C = 256, including Helmholtz's own texts. Two standard modern American textbooks, a 1931 standard phonetics text and the official 1944 physics manual of the U.S. War Department, begin with the standard definition of musical pitch as C = 256.

Regarding composers, all *early music* scholars agree that Mozart tuned at precisely C = 256, as his A was in the range of A = 427–430. Christopher Hogwood, Roger Norrington, and dozens of other directors of orginal-instrument orchestras established the practice during the 1980s of recording all Mozart's works at precisely A = 430, as well as most of Beethoven's symphonies and piano concertos. Hogwood, Norrington, and others have stated, in dozens of interviews and record jackets, the pragmatic reason: German instruments of the period 1780–1827, and even replicas of those instruments, can only be tuned at A = 430.

The demand by Russian Czar Alexander, at the 1815 Congress of Vienna, for a "brighter" sound began the demand for a higher pitch from all the crowned heads of Europe. While classical musicians resisted, the Romantic school, led by Franz Lietz and his father-in-law Richard Wagner, championed the higher pitch during the 1830s and 1840s. Wagner even had the bassoon and many other instruments redesigned so as to be able to play only at A = 440 and higher. By 1850, chaos reigned, with major European theaters at pitches varying from A = 420 to A = 460, and even higher at Venice.

In the late 1850s, the French government, under the influence of a committee of composers led by bel canto proponent Giacomo Rossini, called for the first standardization of the pitch in modern times. France consequently passed a law in 1859 establishing A at 435, the lowest of the ranges of pitches (from A = 434 to A = 456) then in common use in France and the highest possible pitch at which the soprano register shifts may be maintained close to their disposition at C = 256R. It was this French A to which Verdi later referred in objecting to higher tunings then prevalent in Italy, under which circumstance we call an A in Rome what is a B-flat in Paris.

Following Verdi's 1884 efforts to institutionalize A = 432 in Italy, a British-dominated conference in Vienna in 1885 ruled that no such pitch could be standardized. The French, the New York Metropolitan Opera, and many theaters in Europe and the United States continued to maintain their A at 432–435, until World War II.

The first effort to institutionalize A = 440, in fact, was a conference organized in 1939 by Joseph Goebbels, who had standardized A = 440 as the official German pitch. Professor Robert Dussaut of the National Conservatory of Paris told the French press that "By September 1938, the Accoustic Committee of Radio Berlin requested the British Standard Association to organize a congress in London to adopt internationally the German Radio tuning of 440 periods. This congress did in fact occur in London, a very short time before the war, in May–June 1939. No French composer was invited. The decision to raise the pitch was thus taken without consulting French musicians, and against their will." The Anglo-Nazi agreement, given the outbreak of war, did not last, so that still A = 440 did not stick as a standard pitch.

A second congress in London of the International Standardizing Organization met in October 1953, to again attempt to impose A = 440 internationally. This conference passed such a resolution; again no continental musicians who opposed the rise in pitch were invited, and the resolution was widely ignored. "Professor Dussaut of the Paris Conservatory wrote that British instrument makers catering to the U.S. jazz trade, which played at A = 440 and above, had demanded the higher pitch, and it is shocking that so many musicians, orchestra members, and singers should thus be dependent upon jazz players." A referendum by Professor Dussaut of 23,000 French musicians voted overwhelmingly for A = 432.

As recently as 1971, the European Community passed a recommendation calling for the still nonexistent international pitch standard. The action was reported in "The Pitch Game" in the August 9, 1971 issue of *Time* magazine. The article states that $A = 440$, supposedly the international standard, is widely ignored. *Lower tuning* is common, including in Moscow, *Time* reported, "where orchestras revel in a plushy, warm tone achieved by a larynx-relaxing $A = 435$ cycles, and at a performance in London a few years ago, British church organs were still tuned a half-tone lower, about $A = 425$, than the visiting Vienna Philharmonic, at $A = 450$."

For flute acoustics, Coltman [32] has stated that the flute, when viewed as a positive-feedback oscillator, with a resonance frequency, energy input, and loss mechanisms, displays properties that explain physically how the performer adopts his technique to sound the instrument over a three-octave range.

A flute is blown by directing an air stream to strike the edge of the mouth hole, and the air column oscillates at audio frequency.

1.11 Exercises

1.11.1 Show that an open–closed or closed–open tube has a higher resonance frequency than for two open–open or closed–closed tubes of the same wavelength.

1.11.2 For a tube of length 1 m, compute the fundamental frequency for the open–closed or closed–open type tube on the one hand and on the other the closed–closed or open–open tube. For both cases a sonic speed of 330 m/s can be assumed.

1.11.3 For a steel string of diameter 0.1 mm, density $7{,}000\,\mathrm{kgm^{-3}}$ tied between two points 1 m apart, compute the tension that would be required to produce a fundamental frequency of 440 Hz.

1.11.4 Derive the barometric formula for pressure ratio = density ratio, $p_H/p_{H=0} = \rho_H/\rho_{H=0}$, for isothermal air at $T = 300$ K. *Hint*: Take the difference in pressure as $dp = -\rho g dH$, where $g = 9.81\,\mathrm{ms^{-2}}$ is the gravitational constant.

1.11.5 Compute the acoustic power in kilowatts with *Lighthill's formula* of a subsonic jet where jet velocity $= 200\,(\mathrm{m.s^{-1}})$, jet diameter $= 0.4\,(\mathrm{m})$, density $= 1.26\,(\mathrm{kg.m^{-3}})$, and the sonic speed $= 330\,(\mathrm{m.s^{-1}})$. What will the sound intensity be in decibels if an observer hears the sound from 60 m away?

1.11.6 Discuss the whistle and horn sound characteristics of various pieces of everyday sound-producing equipment that you might come across. You may consult the Internet.

Chapter 2
Monopole, Dipole, and Quadrupole Models

In principle, the propagating sound waves caused by a fluctuating source, when there is no reflection from another wall, is not very different from the noise in a tube and the propagation of the sound wave for the pressure fluctuation, except that the boundary conditions must be matched for a given fluctuating source strength.

Let a disturbance be made at $r = 0$ at time t_1 due to a fluctuating source, which reaches the observer at a distance r_o at time t_2. For the propagating medium (air) at rest, the disturbance propagates at sonic speed c_o. Thus, $t_1 = t - r_o/c_o$. It is evident that in the case of a wave train we may define the disturbance by a characteristic value in the wave, for example, by density, which may be given in terms of a Fourier series. For a fluctuating variable in the wave, it can be shown that for a simple harmonic wave the density fluctuation is approximately given proportional to $\cos[\omega(t - r/c_o)]$, and thus

$$\frac{\partial}{\partial r} = -\frac{1}{c_o}\frac{\partial}{\partial t}.$$

(2.1)

For noise generated due to fluctuations in mass, force, and turbulence, Lighthill [61,62] has shown that the acoustic equation can be derived directly from the standard flow equations of conservation and momentum, and some of the terms may be considered to be due to a fluctuating monopole, dipole, or quadrupole. We will now discuss these in a fairly heuristic manner.

2.1 Fluctuating Monopole

2.1.1 Fluctuating Point Mass Source

The fluctuating mass source being considered here initially has an infinitely small volume. In practice, however, this is not true, but one can always consider a case in which the observer hearing the sound is at a sufficiently large distance in relation to

T. Bose, *Aerodynamic Noise: An Introduction for Physicists and Engineers*,
Springer Aerospace Technology 7, DOI 10.1007/978-1-4614-5019-1_2,
© Springer Science+Business Media New York 2013

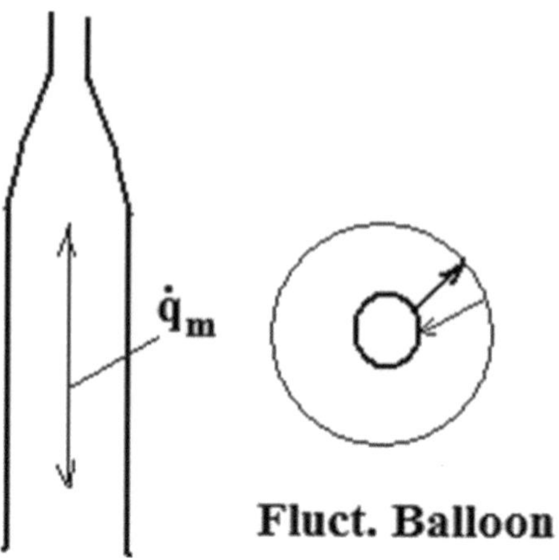

Fig. 2.1 Fluctuating mass source

the characteristic dimension of the source and the wavelength of the radiated sound. Physically, a fluctuating mass source can be generated by several methods (Fig. 2.1). On the left in Fig. 2.1 is a nozzle through which a mass flow rate $\dot{q}_m = \dot{q}_m(t)[\mathrm{kg\,s}^{-1}]$ is flowing. A slightly different arrangement is a *rotating perforated disk* in front of a nozzle, technically known as a *siren*. On the right is an example of a fluctuating balloon, which will also be analyzed later and has the same effect. Note that $\dot{q}_m$ with the dimension of the mass flow rate per unit time is independent of the volume of the actual source.

Thus, let

$t_1 = $ the time at which a sound wave originates at source $(= 0)$,
$t_2 = $ the time at which the sound wave reaches the observer $(= r_o/c_o)$, and
$t = r/c_o$ be the time it takes to travel from the source to the destination.

We consider now a mass flow rate $\dot{q}$ $[\mathrm{kg\,s}^{-1}]$ originating at a point. Let the mass flow rate have a time-averaged part and a time-dependent part, and let the disturbance be fluctuating at a point $r = 0$ at time t_1 and reach the observer at distance r at time t_2. The time taken is $t = t_2 - t_1 = r/c_o$, and we consider now the fluctuating density wave due to the fluctuating mass source. It is evident that in the case of the train of a wave, it is not easy to define a disturbance, but this can be accomplished by arbitrarily fixing a characteristic value in the wave, for example, the maximum value of the density. First, we have a nozzle through which $\dot{q}$ $[\mathrm{kg\,s}^{-1}]$ is flowing, consisting of an average part $\dot{q}$ and a fluctuating part q':

$$\rho(r,t) - \rho_o = \frac{\dot{q}_m(0,t_1)}{4\pi r^2 c_o}. \tag{2.2}$$

From (2.1) and (2.2), we can write

$$\frac{\partial \rho}{\partial r} = -\frac{1}{c_o}\frac{\partial \rho}{\partial t} = -\frac{1}{c_o}\frac{\partial}{\partial t}\left[\frac{\dot{q}_m(0,t_1)}{4\pi r^2}\right]. \tag{2.3}$$

Equation (2.3) can now be integrated if the term within the brackets is considered at r and time $t = r/c_o$ along the path of the integral from $r = 0$ to r:

$$\rho(r,t) = C_1 - \frac{1}{c_o^2} \int_0^r \frac{\partial}{\partial t} \left[\frac{\dot{q}_m(0,t_1)}{4\pi r^2} \right] dr = C_1 + \frac{1}{4\pi c_o^2 r} \frac{\partial \dot{q}_m}{\partial t}, \qquad (2.4)$$

with the boundary condition $\rho(\infty,t) = \rho_o$, and we finally write

$$\rho'(r,t) = \rho(r,t) - \rho_o = \frac{1}{4\pi c_o^2 r} \frac{\partial \dot{q}_m}{\partial t}. \qquad (2.5)$$

The intensity of the sound is now

$$I = \frac{c_o^3}{\rho_o} \overline{\rho'^2} = \left[\frac{1}{\rho_o c_o} \overline{p'^2} \right] = \frac{1}{16\pi^2 r^2 \rho_o c_o} \overline{\left[\frac{\partial \dot{q}_m}{\partial t} \right]^2}, \; \mathrm{Wm}^{-2}, \qquad (2.6)$$

and the acoustic power is obtained by integrating over the surface of a sphere of unit radius, and we get

$$P = \frac{1}{4\pi \rho_o c_o} \overline{\left[\frac{\partial \dot{q}_m}{\partial t} \right]^2}, \; \mathrm{W}. \qquad (2.7)$$

The corresponding relations for pressure and velocity are now

$$p'(r,t) = c_o^2 \rho'(r,t) = \frac{1}{4\pi r} \frac{\partial \dot{q}_m}{\partial t}, \qquad (2.8)$$

$$u' = \frac{c_o^2}{\rho_o} \int \frac{\partial \rho'}{\partial r} dt = \int \frac{\ddot{q}_m}{4\pi r^2} dt = \int \frac{\ddot{\Omega}}{4\pi r^2} dt, \qquad (2.9)$$

where $\dot{\Omega} = \dot{q}_m/\rho_o [\mathrm{m}^3/\mathrm{s}]$ is the *volume flow rate of a source* and $\ddot{\Omega}$ is the rate of change of the volume flow rate of the same source.

Equations (2.8) and (2.9) show further that for $p' \to \infty$ there is a *singularity* as $r \to 0$. This means that to avoid a singularity, a source size of volume $\Omega(t)$ must be considered, and to avoid the singularity we must consider the density and pressure fluctuation away from the *point of singularity*.

2.1.2   *Wave Propagation of a Pulsating Balloon*

In the previous subsection we discussed the case of a fluctuating point mass source, although a point mass source cannot be a practical proposition. A better example would be the fluctuating mass source of a finite dimension, a simple example of which would be a pulsating balloon of radius $R(r)$.

Let the instantaneous volume $V(t)$ and the pulsating balloon of radius $R(t)$.

$$V = V_o + V_1 \exp^{j\omega t},$$

$$R = R_o + R_1 \exp^{j\omega t},$$

where V_o and R_o are respectively the mean volume and radius.

Since

$$V = 4\pi R^2 = V_o + V_1 \exp^{j\omega t}$$

$$= 4\pi \left[R_o + R_1 \exp^{j\omega t} \right]^2$$

$$= 4\pi \left[R_o^2 + 2R_o R_1 \exp^{j\omega t} + R_1^2 \exp^{2j\omega t} \right],$$

therefore

$$V_o = 4\pi R_o^2,$$

$$V_1 = 4\pi \left[2R_o R_1 + R_1^2 \exp^{j\omega t} \right].$$

By making the first derivative of volume and radius with respect to time we obtain

$$\dot{V} = j\omega V_1 \exp^{j\omega t}$$

$$= 8\pi R \dot{R},$$

$$\dot{R} = j\omega R_1 \exp^{j\omega t}.$$

Therefore,

$$\dot{V} = V_1 j\omega \exp^{j\omega t}$$

$$= 8\pi \left[R_o + R_1 \exp^{j\omega t} \right] j\omega R_1 \exp^{j\omega t}.$$

Similarly we produce the second derivative of both of them and obtain

$$\ddot{R} = -\omega^2 R_1 \exp^{j\omega t},$$

$$\ddot{V} = -\omega^2 V_1 \exp^{j\omega t}$$

$$= 8\pi \left[\dot{R}^2 + R\ddot{R} \right]$$

$$= 8\pi \left[-\omega^2 R_1^2 \exp^{2j\omega t} - \omega^2 \exp^{j\omega t} \left(R_o + R_1 \exp^{j\omega t} \right) \right].$$

At the mean radius R_o there will be an equivalent particle velocity on the sphere surface

$$u'_{R_o} = \frac{1}{4\pi R_o^2} \int \ddot{V} dt = \frac{\dot{V}}{4\pi R_o^2}$$

$$= \frac{1}{16\pi R_o^2} \left[R_o + R_1 \exp^{j\omega t} \right] j\omega R_1 \exp^{j\omega t}$$

$$= \frac{1}{16\pi} \left[\left(\frac{R_1}{R_o} \right) + \left(\frac{R_1}{R_o} \right)^2 \exp^{j\omega t} \right] j\omega \exp^{j\omega t}.$$

As long as $R_1/R_o \ll 1$, the second term within [ ] can be neglected, and a simple harmonic wave change in the volume of a completely flexible wall sphere corresponds to a simple harmonic change in the particle velocity of the sphere. In addition, the particle velocity u' on the sphere is in phase with Ω, the rate of change in the volume of the sphere.

Now, the intensity ofthe sound in units of the energy flux rate (Wm^{-3}) is

$$I[\text{W/m}^2] = \frac{c_o^3}{\rho_o}\overline{\rho'^2} = \frac{1}{\rho_o c_o}\overline{p'^2} = \frac{1}{\rho_o c_o}\frac{\overline{\ddot{q}_m^2}}{16\pi^2 r^2}$$

$$= \frac{\rho_o}{c_o}\frac{\overline{\ddot{\Omega}^2}}{16\pi^2 r^2} = \frac{\rho_o \overline{u'^2_{R_o}} R_o^4}{c_o r^2}. \tag{2.10}$$

From (2.7) and (2.8) we can identify the following rules for a fluctuating monopole:

(a) Sound is produced only if the mass flow rate changes with time.
(b) The intensity of sound and acoustic power depends on the square of the amplitude of fluctuation and not on the frequency of oscillation.
(c) The intensity of sound decreases in inverse proportion to the square of the distance.
(d) The intensity of the sound in units of the energy flux is generally proportional to

$$\frac{1}{\rho_o c_o} = \frac{1}{p_o}\frac{\sqrt{R_m T_o}}{\gamma}, \tag{2.11}$$

where $R_m = R^*/m$ is the gas constant, R^* is the *universal gas constant*, m is the mole mass, and γ is the *isentropic coefficient*.

The foregoing is true as long as the amplitude of the fluctuating mass flow rate is not dependent on the density of the gas, for example in sirens. However, there are instances, for example in a fluctuating balloon, where the amplitude is dependent on the mass density of the surrounding gas. For such a situation the intensity of sound (in units of the energy flux) is proportional to

$$I\left[\frac{\text{W}}{\text{m}^2}\right] \sim \frac{\rho_o}{c_o} = \frac{p_o m}{\sqrt{\gamma}(R^* T_o)^{3/2}}, \tag{2.12}$$

where m is the mole mass of the gas.

This last equation can be considered while explaining the results of an experiment by Rayleigh [7]. He noted that the intensity of sound of a bell vibrating inside a jar diminished as the jar was being evacuated. It is obvious from the preceding equation that for the propagation of sound a gaseous surrounding medium must be present. Further, Rayleigh noted that upon introduction of a light gas like hydrogen, the intensity of the sound was smaller than in air at the same pressure. It can also be noted from the two preceding equations that the intensity of the sound of a siren may be louder on a hot day but that of a ringing bell inside a jar may seem less loud.

2.1.3   Spherical Wave Propagation

In the previous subsection we investigated the case of a pulsating sphere in a very simple manner without considering the manner in which the sound wave was propagated from the sphere. This will now be reexamined by considering the propagation of the wave with a view to applying the equations to the example that will follow.

We now examine the propagation of a simple harmonic spherical wave from the surface of a fluctuating sphere. For this let us consider the pressure fluctuation expression as a function of radius and time as

$$p' = c_o^2 \rho' = \frac{A}{r} \exp^{2\pi \mathrm{j}(vt - r/\lambda)}, \mathrm{j} = \sqrt{-1} \tag{2.13}$$

where A is a constant representing the amplitude of the pressure fluctuation. The preceding expression for pressure fluctuations that are inversely proportional to the radius ensures that the fluctuations will vanish as $r \to \infty$. Thus for the *particle velocity* we write

$$u' = -\frac{c_o^2}{\rho} \int \frac{\partial \rho'}{\partial r} \mathrm{d}t = \frac{A}{2\pi \mathrm{j} v r \rho_o} \left[\frac{1}{r} + \frac{2\pi \mathrm{j}}{\lambda} \right] \exp^{2\pi \mathrm{j}(vt - r/\lambda)}$$

$$= \frac{A}{\rho_o c_o r} \left[1 - \frac{\mathrm{j}\lambda}{2\pi r} \right] \exp^{2\pi \mathrm{j}(vt - r/\lambda)}. \tag{2.14}$$

The preceding equation can be split into two components, which merit a little more attention. The real part of the first component gives the gas particle velocity

$$\mathrm{Real}(u') = \frac{A}{\rho_o c_o r} \cos \left[2\pi \left(vt - \frac{r}{\lambda} \right) \right] \exp^{2\pi \left(vt - \frac{r}{\lambda} \right)}, \tag{2.15}$$

which is inversely proportional to r and is equal to $p'/(\rho_o c_o) = c_o \rho'/\rho_o$; this gives the fluctuation of the volume flux rate or the actual fluctuating gas particle velocity of the 1D radially propagating wave. The second component is inversely proportional to r^2 and is $90°$ out of phase with the pressure. For $r = R_o$, the second component can be thought of as the velocity of the gas attached to and moving with the surface of the pulsating sphere. Its value is significant if it is very close to the source, but it is negligible if $\lambda/(2\pi r) \ll 1$ and does not contribute to the acoustic energy radiated away. This leads us to the concept of *near and far fields*.

From (2.14) and $r = R_o$, let the radial velocity be given by $u' = U\cos(\omega t)$, which in complex terms can be written as $u' = U\exp^{j\omega t}$. By substituting this into (2.14) we write

$$U\exp^{j\omega t} = \frac{A}{\rho_o R_o c_o}\left[1 - \frac{j\lambda}{2\pi R_o}\right]\exp^{2\pi j(vt - R_o/\lambda)}. \tag{2.16}$$

Therefore,

$$A = \frac{U\rho_o c_o R_o}{1 - \frac{j\lambda}{(2\pi R_o)}}\exp^{2\pi j(vt - R_o/\lambda)}, \tag{2.17}$$

which after some manipulation becomes

$$A = \frac{U\rho_o c_o R_o\left(1 + \frac{j\lambda}{2\pi R_o}\right)}{1 + \frac{\lambda^2}{4\pi^2 R_o^2}}\left[\cos\left(\frac{2\pi R_o}{\lambda}\right) + j\sin\left(\frac{2\pi R_o}{\lambda}\right)\right]. \tag{2.18}$$

For the particular case of wave length much larger than R_o

$$\frac{\lambda}{2\pi R_o} \gg 1 : \sin\left(\frac{2\pi R_o}{\lambda}\right) \to 0, \cos\left(\frac{2\pi R_o}{\lambda}\right) \to 1, \tag{2.19}$$

and hence

$$A = \frac{2\pi j U\rho_o c_o R_o^2}{\lambda}. \tag{2.20}$$

Thus for this special case only

$$p' = \frac{2\pi j U\rho_o c_o R_o}{\lambda r}\exp^{2\pi j(vt - r/\lambda)}. \tag{2.21}$$

Now the intensity of the sound is

$$I = \overline{p'^2}/(\rho_o c_o), [\mathrm{Wm}^{-2}]. \tag{2.22}$$

Since

$$\overline{\exp^{2j\varphi t}} = \frac{1}{T}\int_0^T \exp^{2j\varphi t}\,\mathrm{d}t = \frac{1}{T}\int_0^T \sin^2(\varphi t)\,\mathrm{d}t = \frac{1}{2}, \tag{2.23}$$

the equation can be simplified more easily if it is written in trigonometric form and where $T = 2\pi/\varphi$ is the period. Thus,

$$\overline{\exp^{2j\varphi t}} = \frac{1}{T}\int_o^T \sin^2(\varphi t)\,\mathrm{d}t = \frac{1}{2\pi}\left[-\frac{1}{4}\sin(2\varphi t) + \frac{\varphi t}{2}\right]_0^{2\pi/\varphi} = \frac{1}{2}. \tag{2.24}$$

Introducing the parameter

$$K = \frac{\lambda}{(2\pi R_o)} \tag{2.25}$$

and writing for the fluctuating pressure

$$p' = \frac{jU\rho_o c_o}{K}\left(\frac{R_o}{r}\right)^2 \exp^{-\frac{jc_o t}{R_o} - \frac{r}{R_o K}}, \tag{2.26}$$

we obtain that the intensity of sound is

$$I(\mathrm{Wm}^{-2}) = -\frac{U^2\rho_o c_o}{2K^2}\left(\frac{R_o}{r}\right)^2 \exp^{-\frac{2jr}{KR_o}}$$

$$= \pm\frac{U^2\rho_o c_o}{2K^2}\left(\frac{R_o}{r}\right)^2 \left[\cos\left(\frac{2r}{KR_o}\right) - j\sin\left(\frac{2r}{KR_o}\right)\right]. \tag{2.27}$$

Further, with a volumetric source strength $\dot{V} = 4\pi R_o^2 U$ (m^{3}s^{-1}), the expressions for fluctuations of velocity, pressure, and density (valid for $K \gg 1$) are

$$u' = \frac{j\dot{V}}{2\lambda r}\left[1 - \frac{j\lambda}{2\pi r}\exp^{2\pi j(vt - r/\lambda)}\right], \tag{2.28}$$

$$p' = \frac{j\rho_o c_o \dot{V}}{2\lambda r}\exp^{2\pi j(vt - r/\lambda)}, \tag{2.29}$$

$$\rho' = \frac{p'}{c_o^2} = \frac{j\rho_o \dot{V}}{2\lambda c_o r}\exp^{2\pi j(vt - r/\lambda)}. \tag{2.30}$$

The three preceding fluctuations are obviously in phase. However, for $K \ll 1$, the more complete equations for fluctuations of velocity and pressure are

$$u' = U\frac{1 + jK(1 - R_o/r) + K^2\frac{R_o}{r}}{1 + K^2}\exp^{2\pi jvt + jK(1 - r/R_o)}, \tag{2.31}$$

$$p' = \frac{U\rho_o c_o(1 + jK)}{K^2}\frac{R_o}{r}\exp^{\left[\frac{j}{K}\left(\frac{c_o t}{R_o} + 1 - \frac{r}{R_o}\right)\right]}. \tag{2.32}$$

It is evident that, depending on the value of K, u' and p' may or may not be in phase. Further, for an arbitrary value of K, the intensity of sound is

$$I = \frac{U^2\rho_o c_o}{K^4}(1 - K^2 + 2jK)\left(\frac{R_o}{r}\right)^2 \exp^{\left[\frac{2j}{K}\left(1 - \frac{r}{R_o}\right)\right]}, \mathrm{Wm}^{-2}, \tag{2.33}$$

which depends on K, that is, frequency or wavelength, and on r/R_o. This result will become clearer as we discuss sound radiation from a pulsating piston, in which the approximation is made that $K \gg 1$. In all these relations, it is evident that for the far field, as $R_o/r \to 0$, the pressure fluctuation vanishes and the intensity goes to 0.

Fig. 2.2 Pulsating piston

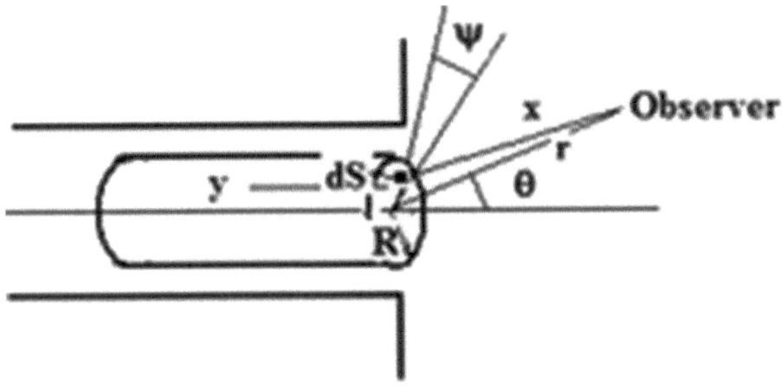

The effect of cutting the source and the infinite space in half by means of a rigid baffle is to leave the pressure field exactly as it was, but this effect is produced only on one side of the baffle by a pulsating sphere of strength $\dot{V} = 2\pi R_o^2 U$. Then the modified pressure field is

$$p' = \frac{j\rho_o c_o \dot{V}}{\lambda r} \exp^{\frac{j}{k}\left[\frac{c_o t}{R_o} - \frac{r}{r_o}\right]}, \tag{2.34}$$

with

$$k = \frac{2\pi}{\lambda}, \quad p' = \frac{j\rho_o c_o k \dot{V}}{2\pi r} \exp^{jk(\omega t - r)}, \tag{2.35}$$

and we consider the case of a radiating pulsating piston (Fig. 2.2).

From an elemental surface dS at a distance $\mathbf{y}$ from the axis, a sound wave radiates to give a pressure fluctuation dp', which, integrated for the contribution from the total surface, gives the total pressure fluctuation p'. Now suppose the whole piston surface is vibrating uniformly at a velocity $U\exp^{j\omega t}$; thus

$$dp' = \frac{j\rho_o c_o k}{2\pi x} \exp^{j(2\pi vt - kx)} dS; \omega = 2\pi v. \tag{2.36}$$

Now d$S = y\,dy\,d\psi$ and

$$x = \sqrt{r^2 + y^2 - 2ry\sin\theta\cos\psi} \approx r - y\sin\theta\cos\psi. \tag{2.37}$$

Therefore, the fluctuating pressure is

$$dp' = \frac{j\rho_o c_o k U}{2\pi r[1 - (y/r)\sin\theta\cos\psi]} \exp^{j(2\pi vt - kr + ky\sin\theta\cos\psi)} y\,dy\,d\psi$$

$$= \frac{j\rho_o c_o k U}{2\pi r} \exp^{j(2\pi vt - kr + ky\sin\theta\cos\psi)} y\,dy\,d\psi,$$

and the total fluctuating pressure is

$$p' = \frac{j\rho_o c_o k U}{2\pi r} \exp^{j(2\pi vt - kr)} \int_{y=0}^{R} \int_{\psi=0}^{2\pi} \exp^{j(ky\sin\theta\cos\psi)} y\,dy\,d\psi. \tag{2.38}$$

Expanding the inner integral in series we obtain

$$p' = \int_{\psi=0}^{2\pi} \exp^{jky\sin\theta\cos\psi} d\psi = 2\pi J_o(ky\sin\theta), \qquad (2.39)$$

where

$$J_o(\psi) = 1 - \frac{\psi^2}{2^2} + \frac{\psi^4}{2^2 4^2} - + \cdots \qquad (2.40)$$

is the Bessel function of zeroth order of φ. This is obtained in the following manner.
Since

$$\exp^\psi = 1 + \frac{\psi}{1!} + \frac{\psi^2}{2!} + \frac{\psi^3}{3!} + \cdots,$$

$$\exp^{jky\sin\theta\cos\psi} = 1 + \frac{jky\sin\theta\cos\psi}{1!} - \frac{(ky\sin\theta\cos\psi)^2}{2!} + - \cdots.$$

Further, since

$$\int \cos^n \psi d\psi = \frac{\sin\psi\cos^{n-1}\psi}{n} + \frac{n-1}{n} \int \cos^{n-2}\psi d\psi, \qquad (2.41)$$

then

$$\int_{\psi=0}^{2\pi} \exp^{jky\sin\theta\cos\psi} d\psi = \int_{\psi=0}^{2\pi} \left[1 + \sum_{n=1}^{\infty} \frac{(jk_y\sin\theta\cos\psi)^n}{n!} \right] d\psi$$

$$= 2\pi + \sum_{n=1}^{\infty} \frac{(jky\sin\theta)^n}{n!} \int_{\psi=0}^{2\pi} \cos^n\psi d\psi$$

$$= 2\pi J_o(ky\sin\theta).$$

Therefore,

$$p' = \frac{j\rho_o c_o kU}{2\pi} \exp^{j(2\pi vt - kr)} \int_{y=0}^{R} 2\pi J_o(ky\sin\theta)ydy$$

$$= \frac{j\rho_o c_o kU \pi R^2}{2\pi r} \exp^{j(2\pi vt - kr)} \left[\frac{2J_1(kr\sin\theta)}{kR\sin\theta} \right], \qquad (2.42)$$

where J_1 is the *Bessel function of first order*,

$$J_1 = \frac{1}{2} \left[\varphi - \frac{\varphi^3}{2\times4} + \frac{\varphi^5}{2\times4\times4\times6} \mp \cdots \right]. \qquad (2.43)$$

Further, the term between the [ ] is the directivity factor for acoustic radiation of
the pulsating piston. It can be shown that $F(\theta)$ is zero for some values of $\sin\theta$, and
there is no radiation in that direction (Fig. 2.3). This was shown in Fig. 2.4.

As a result, the intensity distribution for a pulsating piston is shown in Fig. 2.3.

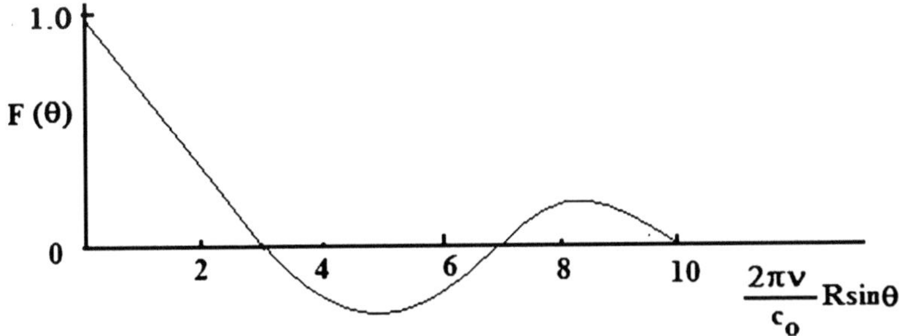

Fig. 2.3 Directivity factor for acoustic radiation of the pulsating piston

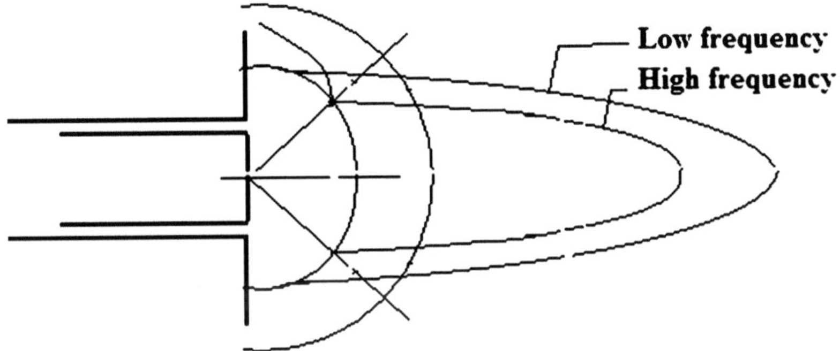

Fig. 2.4 Intensity distribution for a pulsating piston

2.2   Fluctuating Dipole

In the previous section we described how we introduced fluid and withdrew it periodically from a small region in space, the source point, to generate sound (monopole). Although the condition may be fulfilled in many cases, for dipoles, these are produced by moving a portion of the fluid as the source or sink at a neighboring point but not introducing any new fluid. The simplest arrangement of this sort can be simulated by two simple sources completely out of phase with each other (equal in magnitude but with opposite signs). Such a source is called as a *dipole source.*

The physical model of a fluctuating dipole can be thought of as two balloons connected by a rigid small tube of length L in which while one balloon contracts, the other expands, and vice versa. Thus we have two fluctuating monopoles completely in phase opposite to each other, shown symbolically in Fig. 2.5a. Alternatively, we can consider two faces of a reciprocating piston (Fig. 2.5b).

In both these cases we have a mathematical model of a dipole arising out of alternating fluctuating monopole sources (Fig. 2.6).

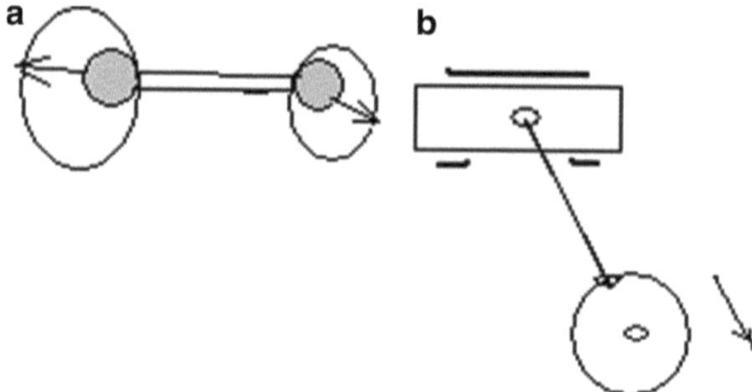

Fig. 2.5 Fluctuating dipole sources

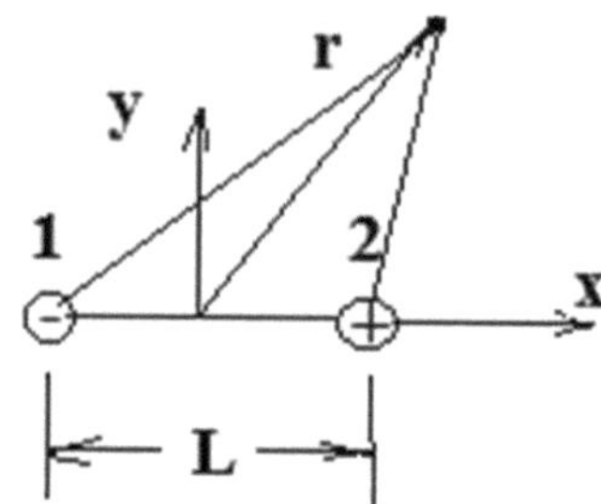

Fig. 2.6 Mathematical model of fluctuating dipole source

On the basis of geometrical considerations,

$$r_{1,2} = \sqrt{(x \pm L/2)^2 + y^2 + z^2}. \tag{2.44}$$

With $x/r = \cos\theta$ and $x^2 + y^2 + z^2$, we can rewrite the preceding expressions as

$$r_{1,2} = r\sqrt{1 \pm (L/r)\cos\theta + (L/r)^2} \approx r\sqrt{1 \pm (L/r)\cos\theta}. \tag{2.45}$$

If t is the time at which the signals originating at 1 and 2 reach the observer, then

$$t_1 = t - r_1/c_o \text{ and } t_2 = t - r_2/c_o, \tag{2.46}$$

and thus the time lag in the origination of the two signals from 1 and 2, which reach the observer simultaneously, is

$$\tau = t_1 - t_2 = \frac{|r_1 - r_2|}{c_o}$$

$$= \frac{r}{c_o}\left[\sqrt{1 - \left(\frac{L}{r}\right)\cos\theta + \left(\frac{L}{2r}\right)^2} - \sqrt{1 + \left(\frac{L}{r}\right)\cos\theta + \left(\frac{L}{2r}\right)^2}\right]$$

$$\approx \frac{r}{c_o}\frac{L}{r}\cos\theta = \frac{L}{c_o}\cos\theta. \tag{2.47}$$

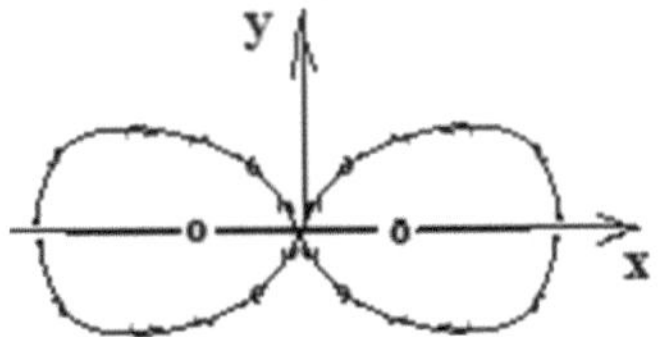

Fig. 2.7 Radiation directivity of a dipole

If it is found that $|\tau| > \ll T = v^{-1}$, the characteristic period of oscillations, then it is possible to neglect the time delay. Under this assumption,

$$\rho' = \rho(r,t) - \rho_o = \frac{\dot{q}_m(0, t - r_2/c_o)}{4\pi r_2^2 c_o} - \frac{\dot{q}_m(0, t - r_1/c_o)}{4\pi r_1^2 c_o}$$

$$= \frac{\ddot{q}_m}{4\pi c_o^2 r}\left[\frac{1}{1 - \frac{L}{r}\cos\theta} - \frac{1}{1 + \frac{L}{r}\cos\theta}\right] = \frac{\ddot{q}_m L\cos\theta}{4\pi c_o^2 r^2} \tag{2.48}$$

for $L/r \ll 1$. Using a method similar to that used in the case of a monopole, we can now write after integration

$$\rho' = \rho(r,t) - \rho_o = \frac{\ddot{q}_m(L/r)\cos\theta}{8\pi c_o^2 r}, \text{kgm}^{-3}, \tag{2.49}$$

where $\ddot{q}_m$ is in kilograms per square second.

Thus the intensity of sound is given by the relation

$$I = \frac{c_o^3}{\rho_o}\overline{(\rho')^2} = \frac{\overline{\ddot{q}_m^2}(L/r)^2\cos^2\theta}{64\pi^2 c_o\rho_o r^2}, \text{Wm}^{-2}. \tag{2.50}$$

Obviously for $\theta = 90°, I = 0$ and at $\theta = 0$ or $180°$, the maximum intensity is

$$I_{\max} = \frac{\overline{\ddot{q}_m^2}(L/r)^2}{64\pi^2 c_o\rho_o r^2}, \text{Wm}^{-2}. \tag{2.51}$$

Therefore,

$$\frac{I}{I_{\max}} = \cos^2\theta, \tag{2.52}$$

and the result is plotted in Fig. 2.7.

Hence the *acoustic power of a single dipole* is

$$P = \frac{\overline{\ddot{q}_m^2}(L/r)^2}{16\pi c_o\rho_o}, \text{W}, \tag{2.53}$$

where in (L/r), $r = 1m$, and L is also in m.

While the foregoing analysis gives the angular distribution of the dipole intensity, we will now examine the *cancellation effects in dipole* of the distance between the two monopoles that constitute a dipole. For this purpose we consider the effect on

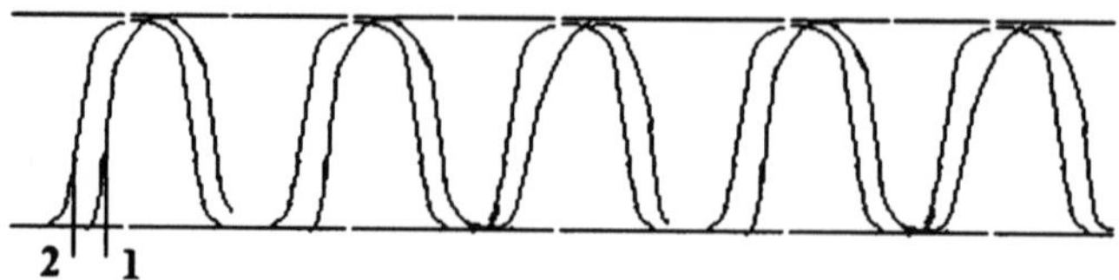

Fig. 2.8 Propagation of waves due to two fluctuating sources being in phase

the line connecting the two monopoles. The density and pressure wave propagation due to a mass source has an associated equivalent force, $f = \dot{q}_m c_o$. In the 1D case, the total of the two equivalent forces, due to propagation of waves from two fluctuating sources (in phase), are (Fig. 2.8)

$$f_{1,2} = Ac_o \cos\left[\omega\left(t - \frac{x \pm L/2}{c_o}\right)\right], \tag{2.54}$$

as shown in Fig. 2.9, and the total force is

$$F = f_1 + f_2 = 2Ac_o \cos\left[\omega\left(t - \frac{x}{c_o}\right)\right]\cos\left(\frac{\pi L}{\lambda}\right), \tag{2.55}$$

which becomes maximum in the limit $L \to 0$,

$$\lim_{L \to 0}(F) = 2Ac_o \cos\left[\omega\left(t - \frac{x}{c_o}\right)\right]. \tag{2.56}$$

Thus both the sources are in phase and have no distance between them, and the total mass flow rate is given by

$$\dot{q}_m = 2A \cos\left[\omega\left(t - \frac{x}{c_o}\right)\right]. \tag{2.57}$$

On the other hand, the two monopoles in a dipole must be completely out of phase, and thus the dipole source strength is obtained by subtracting the two forces, and we get

$$f = f_1 - f_2 = 2Ac_o \sin\left[\omega\left(t - \frac{x}{c_o}\right)\right]\sin\left(\frac{\pi L}{\lambda}\right). \tag{2.58}$$

The first time derivative terms of the mass flow rate with respect to t and x are

$$\frac{\partial \dot{q}_m}{\partial t} = 2A\omega \sin\left[\omega\left(t - \frac{x}{c_o}\right)\right], \tag{2.59}$$

$$\frac{\partial \dot{q}_m}{\partial x} = -2A\omega \sin\left[\omega\left(t - \frac{x}{c_o}\right)\right]\sin\left(\frac{\pi L}{\lambda}\right), \tag{2.60}$$

and the relative magnitude of the noise-producing term is

$$\varepsilon_1 = \left| \frac{\partial f/\partial x}{\partial \dot{q}_m/\partial t} \right| = \sin\left(\frac{\pi L}{\lambda}\right), \tag{2.61}$$

which means that

$$L < \frac{\lambda}{2} : \varepsilon_1 < 1; \quad L = 0 \quad \text{or} \quad \lambda : \varepsilon = 0, \tag{2.62}$$

and thus it is shown that a dipole system is less efficient than the equivalent monopole system since in a dipole system the fluctuating sources cancel each other to some extent. If $L = 0$ or $L = \lambda \to \varepsilon_1 = 0$, then the cancellation of these two sources is complete for these two values of L.

We showed in (2.5) that the noise-producing term for a fluctuating mass source causes the density fluctuation

$$\rho'(r,t) = \frac{1}{4\pi r c_o^2} \frac{\partial \dot{q}_m}{\partial t}. \tag{2.63}$$

Now, at least qualitatively, the noise-producing term $\ddot{q}_m$ can be replaced by an equivalent term $\partial f/\partial x$, multiplied by a factor $\sin(\pi L/\lambda)$, and thus we write in an analogous manner

$$\rho(r,t) - \rho_o = \frac{\ddot{q}_m}{4\pi c_o^2 r}. \tag{2.64}$$

Further,

$$\frac{1}{r} \frac{\partial \dot{q}_m}{\partial t} \equiv \frac{\partial}{\partial t}\left(\frac{\dot{q}_m}{r}\right) \equiv \frac{\partial}{\partial x}\left(\frac{c_o \dot{q}_m}{r}\right) \equiv \frac{\partial}{\partial x}\left(\frac{f}{r}\right). \tag{2.65}$$

Hence,

$$\rho'(r,t) = \frac{1}{4\pi c_o^2} \frac{\partial}{\partial x_i}\left(\frac{f_i}{r}\right). \tag{2.66}$$

Since $r^2 = \sum x_i^2$ and $r dr = x_i dx_i$, we get

$$\begin{aligned}
\rho' &= \frac{1}{4\pi c_o^2}\left[\frac{1}{r}\frac{\partial f_i}{\partial x_i} + f_i \frac{\partial}{\partial x_i}\left(\frac{1}{r}\right)\right] = \frac{1}{4\pi c_o^2}\left[\frac{1}{r}\frac{\partial f_i}{\partial r}\frac{\partial r}{\partial x_i} + f_i \frac{\partial}{\partial x_i}\left(\frac{1}{r}\right)\right] \\
&= \frac{1}{4\pi c_o^2}\left[\frac{x_i}{r^2}\frac{\partial f_i}{\partial r} - \frac{x_i f_i}{r^3}\right] = \frac{1}{4\pi c_o^2}\left[-\frac{x_i}{c_o r^2}\dot{f}_i - \frac{x_i f_i}{r^3}\right].
\end{aligned} \tag{2.67}$$

In the preceding equation, obviously the quantities within brackets must be evaluated at a retarded time $(t - r/c_o)$, and then

$$\rho'(r,t) = \frac{1}{4\pi c_o^2}\left[-\frac{x_i}{c_o r^2}\dot{f}\left(0, t - \frac{r}{c_o}\right) - \frac{x_i}{r^3}f_i\left(0, t - \frac{r}{c_o}\right)\right]. \tag{2.68}$$

Please note that the first term within [ ] is the time derivative term required for production of sound wave. Assuming a simple harmonic wave form of f_i

$$f_i = 2A[\text{kg s}^{-1}]c_o \sin\left[2\pi v\left(t - \frac{r}{c_o}\right)\right] \sin\left(\frac{\pi L}{\lambda}\right) \tag{2.69}$$

we can write

$$\rho'(r,t) = \frac{2A}{4\pi c_o} \sin\left(\frac{\pi L}{\lambda}\right)$$

$$\times \frac{x_i}{r^3}\left[\frac{2\pi r}{\lambda}\cos\left(2\pi v\left(t - \frac{r}{c_o}\right)\right) + \sin\left(2\pi v\left(t - \frac{r}{c_o}\right)\right)\right]. \tag{2.70}$$

Taking the root mean square of the two terms within [ ] we can show that for

$$\varepsilon_2 = \frac{\lambda}{2\pi r} \ll 1 \tag{2.71}$$

which is for the far field approximation, when the second term within [ ] can be neglected. Hence it is concluded that between the two terms in (2.70), the first term refers to the far field and the second term to the near field. In addition (2.70) not only depends on the magnitude of the dipole source strength, but on the distance L between the two monopoles of dipole, when the term becomes zero for $L = 0$ or $L = 2\lambda$.

But monopoles and dipoles are not the only two sound-producing mechanisms. We will show in the next section how a quadrupole, as a combination of two dipoles, can be another mechanism to produce sound.

2.3 Fluctuating Quadrupole

The physical mechanical model, as a combination of two dipoles, is shown in Fig. 2.9. While in Fig. 2.9a the two dipoles are set parallel to each other (lateral quadrupole), in Fig. 2.9b they are put in the same line (longitudinal quadrupole).

For a lateral quadrupole we consider a combination of four source sinks, as shown in Fig. 2.10. Now,

$$x/r = \cos\theta; y/r = \sin\theta; r^2 = x^2 + y^2 + z^2. \tag{2.72}$$

Further, the distance between the observer and each of the source sinks is

$$r_k = \sqrt{(x \pm L_x/2)^2 + (y \pm L_y/2)^2 + z^2} \approx r\sqrt{1 \pm \left(\frac{L_x\cos\theta \pm L_y\sin\theta}{r}\right)}. \tag{2.73}$$

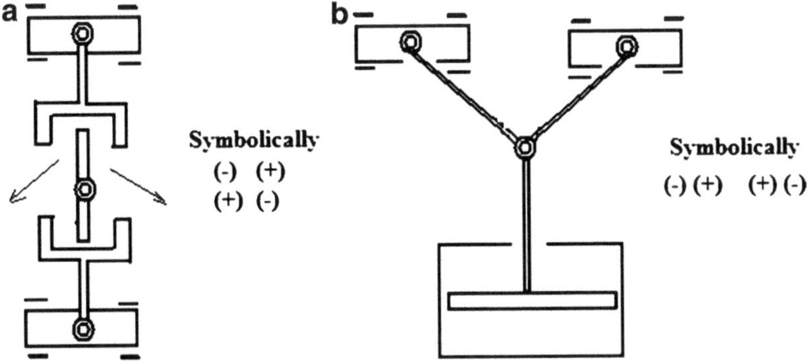

Fig. 2.9 Models of fluctuating quadrupole

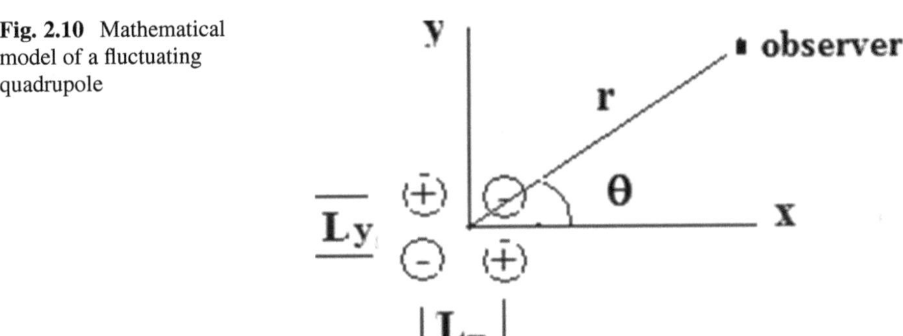

Fig. 2.10 Mathematical model of a fluctuating quadrupole

The time delay between signals from any two sources to the observer is

$$|\Delta t_{ij}| = \frac{|r_i - r_j|}{c_o}. \tag{2.74}$$

It can be shown again, as in the case of the dipole, that for frequencies less than 20,000 cycles/s (Hz), different parts of a quadrupole can be added. Now the density fluctuation is

$$\rho'(r,t) = \frac{1}{4\pi c_o} \sum_{i=1}^{4} \frac{\dot{q}_{mi}(0, t - r/c_o)}{r_i^2}$$

$$\approx \frac{\dot{q}_m L_x L_y \sin(2\theta)}{\pi c_o r^4}, \tag{2.75}$$

where for $L_x/r \ll 1$ and $L_y/r \ll 1$,

$$\rho'(r,t) = \frac{\dot{q}_m(0,t-r/c_o)}{\pi c_o r^4} L_x L_y \sin(2\theta), \left[\frac{\text{kg}}{\text{m}^3}\right]. \tag{2.76}$$

After integration, as was done for the fluctuating dipole, we get

$$\rho'(r,t) = \frac{\ddot{q}_m(0,t-r/c_o)}{3\pi c_o^2 r} \frac{L_x}{r} \frac{L_y}{r} \sin 2\theta. \tag{2.77}$$

The expression for the intensity of the sound is now

$$I = \frac{\left(\frac{L_x}{r}\frac{L_y}{r}\right)^2}{9\pi^2 c_o r^2} \overline{\ddot{q}_m}^2 \sin^2 2\theta. \, \text{Wm}^{-2}, \tag{2.78}$$

from which it can be seen that for a *fluctuating lateral quadrupole* the intensity of sound is

$$I \propto \sin^2(2\theta) \propto \sin^2(\theta)\cos^2(\theta). \tag{2.79}$$

The *acoustic power of the fluctuating lateral quadrupole* is now

$$P = \frac{4\left(\frac{L_x}{r}\frac{L_y}{r}\right)^2}{9\pi c_o \rho_o} \overline{\ddot{q}_m}^2. \, \text{W}. \tag{2.80}$$

In evaluating the preceding equation, r is to be $1m$, and L_x and L_y must also be in m.

From (2.79) it can be seen, therefore, that the intensity of a sound from a fluctuating lateral quadrupole is zero at $\theta = 0°$, $90°$, $180°$, and $270°$, and the maximum is at $\theta = 45°$, $135°$, $225°$, and $315°$. These results are plotted in Fig. 2.11.

Now similar considerations are made for a fluctuating longitudinal quadrupole, a mathematical model of which is shown in Fig. 2.12.

Here, since $x/r = \cos\theta$, $y/r = \sin\theta$, and $x^2 + y^2 + z^2 = r^2$, the distances from individual source sinks to the observer are

$$r_{4,1} = \sqrt{(x \pm (L' \pm L/2)^2 + y^2 + z^2} \approx r\left[1 \pm \frac{(2L'+L)\cos\theta}{2r}\right] \tag{2.81}$$

and

$$r_{3,2} = \sqrt{(x \pm L/2)^2 + y^2 + z^2} \approx r\left[1 \pm \frac{L\cos\theta}{2r}\right]. \tag{2.82}$$

The time delay between signals from any two sources to the observer is $|\Delta_{ij}| = |r_i - r_j|/c_o$, which can be written in matrix form as follows (Table 2.1).

Now the maximum time delay is between 1 and 4:

$$|\Delta t_{\max}| = \frac{|r_1 - r_4|}{c_o} = \frac{|(2L'+L)\cos\theta|}{c_o}, \tag{2.83}$$

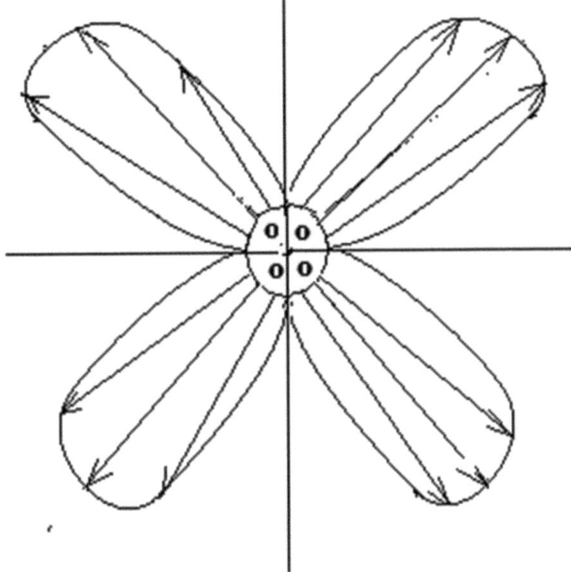

Fig. 2.11 Angular intensity distribution of a lateral quadrupole

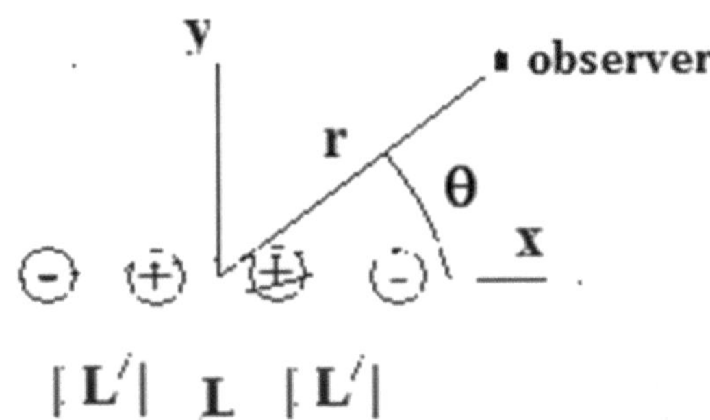

Fig. 2.12 Mathematical model of a fluctuating quadrupole

Table 2.1 Time delay $\Delta\tau_{ij}$ between signals of two sources

	1	2	3	4
1	0	$\dfrac{\lvert L_y\sin\theta\rvert}{c_o}$	$\dfrac{\lvert L_x\cos\theta+L_y\sin\theta\rvert}{c_o}$	$\dfrac{\lvert L_x\cos\theta\rvert}{c_o}$
2	$\dfrac{\lvert L_y\sin\theta\rvert}{c_o}$	0	$\dfrac{\lvert L_x\cos\theta\rvert}{c_o}$	$\dfrac{\lvert L_x\cos\theta-L_y\sin\theta\rvert}{c_o}$
3	$\dfrac{\lvert L_x\cos\theta+L_y\sin\theta\rvert}{c_o}$	$\dfrac{\lvert L_x\cos\theta\rvert}{c_o}$	0	$\dfrac{\lvert L_y\sin\theta\rvert}{c_o}$
4	$\dfrac{\lvert L_x\cos\theta\rvert}{c_o}$	$\dfrac{\lvert L_x\cos\theta-L_y\sin\theta\rvert}{c_o}$	$\dfrac{\lvert L_y\sin\theta\rvert}{c_o}$	0

which can be neglected if $\lvert\Delta t_{\max}\rvert\nu\ll 1$. It is found that generally the condition is met for frequencies less than 20,000 cycles/s.

As in the case of a fluctuating lateral quadrupole, we can write for the longitudinal quadrupole the density function as

$$\rho(r,t) - \rho_o = \sum_{i=1}^{\infty} \frac{\dot{q}_{mi}(0.t - r/c_o)}{4\pi c_o r_i^2}$$

$$= \frac{4\dot{q}_m \cos^2\theta (L+L')L'/2}{\pi c_o r^2 \left(1 - \frac{2L'+L)^2}{r^2}\cos^2\theta\right) \frac{1}{\left(1-\left(\frac{L}{r}\right)^2 \cos^2\theta\right)}}. \tag{2.84}$$

Integrating as previously, and for $L = L' \ll r$, we get

$$\rho(r,t) - \rho_o = \frac{4}{3\pi} \frac{\ddot{q}_m}{c_o^2 r}\left(\frac{L}{r}\right)^2 \cos^2\theta. \tag{2.85}$$

The *intensity of sound for a longitudinal quadrupole* is now

$$I = \frac{16}{9\pi^2} \frac{\overline{\ddot{q}_m^2}}{\rho_o c_o r^2}\left(\frac{L}{r}\right)^2 \cos^4\theta \tag{2.86}$$

and the *acoustic power of a longitudinal quadrupole* is

$$P = \frac{64}{9\pi} \frac{\overline{\ddot{q}_m^2}}{\rho_o c_o}. \tag{2.87}$$

Thus the intensity I (W/m^{-2}) is proportional to $\cos^4\theta$, whereas it was shown earlier for a fluctuating dipole that the intensity distribution was proportional to $\cos^2\theta$; the result of the latter was shown in Fig. 2.7. In the case of fluctuating longitudinal quadrupole, the results indicates (not shown here explicitly) that the intensity distribution profile is comparatively slender in comparison to that given in Fig. 2.7.

Thus it is found that the directivity patterns of elementary sources and sinks are as follows:

$$\text{monopole} \sim 1, \tag{2.88}$$

$$\text{dipole} \sim \cos^2\theta, \tag{2.89}$$

$$\text{lateral quadrupole} \sim \sin^2\theta \cos^2\theta, \tag{2.90}$$

$$\text{longitudinal quadrupole} \sim \cos^4\theta. \tag{2.91}$$

The formulation of a quadrupole as a combination of two dipoles (lateral or longitudinal) is not quite how Lighthill discussed the analogy between certain terms in fluid dynamic equations and quadrupoles and dipoles, as we will discuss in the next chapter. According to Lighthill, the analogous quantity of a quadrupole is

$f_{ij} = \rho u_i u_j$ (kgm^{-1}s^{-2}) and the analogous quadrupole strength for fluctuations in density is proportional to $\partial^2 f_{ij}/\partial x_i \partial x_j$. The proportionality factor is the inverse of $4\pi c_o^2$.

Since $\sum x_i^2 = r^2$, a further derivation is

$$\rho - \rho_o = \frac{1}{4\pi c_o^2} \frac{\partial^2}{\partial x_i \partial x_j} \left(\frac{\partial^2 f_{ij}}{\partial r^2} \right) = \frac{x_i x_j}{4\pi c_o^2 r^3} \left[\frac{1}{c_o^2 r} \ddot{f}_{ij} + \frac{3}{c_o r} \dot{f}_{ij} + \frac{f_{ij}}{r^2} \right], \qquad (2.92)$$

where the first term will dominate in comparison to the other terms if $\lambda/(2\pi r) \ll 1$ (far field!).

For experimental measurements of radiated sound, it is now possible to make an estimate of the distance at which the measured sound can be considered to be in the far fields. Since a human ear can hear sound at frequencies of a few cycles per second (say, 10 cycles per second) to about 20,000 cycles per second, the corresponding wavelength range will be 0.0165 to 33 m if the ground speed is 330 ms^{-1}. Assuming a maximum value of $\lambda/(2\pi r)$ to be 0.01 in the far field, the far field distance is 525 m at 10 cycles per second to 0.2625 m at 20,000 cycles per second. Therefore, it is estimated that for a comparatively low-frequency noise that occurs in piston engines, an observer is probably never sufficiently in the far field, but for turbulent jet noise, it is estimated that a distance of 1 m may be sufficient to consider the measurements as far field noise.

2.4 Exercises

2.4.1 Explain physically the monopole, dipole, and quadrupole radiations. What is the difference between lateral and longitudinal quadrupoles? For heat radiation, there is no radiation for homopolar diatomic molecules such as, for example, H_2, O_2, and N_2, whereas for heteropolar diatomic molecules such as, for example, OH and NO there can be heat radiation. Explain.

2.4.2 For a fluctuating monopole, if 1 (kg s^{-1}) of air on average is discharged at a frequency of 1,000 Hz and an observer is at a distance of 20 m, calculate (a) the radian frequency ω, the wavelength λ, the amplitude of the mass flow rate, and the first time derivative of the mass flow rate $\ddot{q}_m$; (b) the intensity of sound in (W m^{-2}) and in decibels; and (c) the acoustic power (W).

2.4.3 For a single dipole consisting of two monopoles with a mass flow rate of 0.5 (kgs^{-1}), each 1 cm apart, calculate (a) the intensity of sound for an observer at a distance of 20 m and making an angle (in degrees) of 0, 45, and 90 in (W m^{-2}) and in decibels for each of the angles; and (b) the acoustic power (W). How does it compare with the acoustic power of the monopole in the previous example?

Chapter 3
Lighthill's Theory of Aerodynamic Noise

In 1952, Lighthill [61] gave the first formulation of aerodynamic sound, which is based on an analogy of sound radiated by fluctuating monopole, dipole, or quadrupole sound sources. Lighthill's treatment of the subject will be discussed in this chapter, which starts with a derivation of an equation of sound.

3.1 Lighthill's Equation of Sound

Lighthill's equation of sound is derived from the fundamental equations of fluid mechanics, namely, the equations of continuity and momentum. These equations are written below.

Continuity:

$$\frac{\partial \rho}{\partial t} + \sum_j \frac{\partial(\rho u_j)}{\partial x_j} = \dot{Q}_m, \ \text{kgm}^{-3}\text{s}^{-1};$$ (3.1)

Momentum:

$$\frac{\partial(\rho u_i)}{\partial t} + \sum_j \frac{\partial(\rho u_i u_i u_j)}{\partial x_j} = \sum_j \frac{\partial \tau_{ij}}{\partial x_j} + F_i, \ \text{Nm}^{-3},$$ (3.2)

where the shear stress is given by

$$\tau_{ij} = \left(-p - \frac{2}{3}\mu \sum \frac{\partial u_j}{\partial x_j}\right) \delta_{ij} + \mu \left(\frac{\partial u_i}{\partial x_j} + \frac{\partial u_j}{\partial x_i}\right) = p\delta_{ij} - \tau_{ij}^*,$$

$$\delta_{ij} = \text{Kronecker delta}: \delta_{ii} = 1, \delta_{ij} = 0,$$

$$\tau_{ij}^* = \left(\frac{2}{3}\mu \sum \frac{\partial u_j}{\partial x_j}\right) \delta_{ij} - \mu \left(\frac{\partial u_i}{\partial x_j} + \frac{\partial u_j}{\partial x_i}\right).$$ (3.3)

T. Bose, *Aerodynamic Noise: An Introduction for Physicists and Engineers*,
Springer Aerospace Technology 7, DOI 10.1007/978-1-4614-5019-1_3,
© Springer Science+Business Media New York 2013

Thus (3.2) is rewritten as

$$\frac{\partial(\rho u_i)}{\partial t} + \sum_j \frac{\partial(\rho u_i u_j)}{\partial x_j} = -\frac{\partial p}{\partial x_j} - \sum_j \frac{\partial \tau_{ij}^*}{\partial x_j} + F_i, \, \mathrm{Nm}^{-3}. \tag{3.4}$$

Differentiating (3.1) with respect to t and the three-momentum equation (3.4) with respect to the corresponding x_i and adding we get

$$\frac{\partial^2 \rho}{\partial t^2} = -\sum_j \frac{\partial^2(\rho u_i)}{\partial t \partial x_i} + \frac{\partial \dot{Q}_m}{\partial t}, \tag{3.5}$$

$$\sum_i \frac{\partial^2 p}{\partial x_i^2} = \sum_i \frac{\partial^2(\rho u_i)}{\partial t \partial x_i} - \sum_i \sum_j \frac{\partial(\rho u_i u_j)}{\partial x_i \partial x_j} - \sum_i \sum_j \frac{\partial^2 \tau_{ij}^*}{\partial x_i \partial x_j} + \sum_i \frac{\partial F_i}{\partial x_i}. \tag{3.6}$$

Subtracting the second equation from the first we get

$$\frac{\partial^2 \rho}{\partial t^2} - \sum_i \frac{\partial^2 p}{\partial x_i^2} = \frac{\partial \dot{Q}_m}{\partial t} + \sum_i \sum_j \frac{\partial^2 T_{ij}^*}{\partial x_i \partial x_j} - \sum_i \frac{\partial F_i}{\partial x_i} = \dot{S}, \, \mathrm{kgm}^{-3}\mathrm{s}^{-1}, \tag{3.7}$$

where

$$T_{ij}^* = \tau_{ij}^* + \rho u_i u_j = \tau_{ij}^* + T_{ij}, \tag{3.8}$$

where the quadrupole moment

$$T_{ij} = \rho u_i u_j, \, \mathrm{kgm}^{-1}\mathrm{s}^{-2} \tag{3.9}$$

Comparing (3.8) and (3.9) we can obviously write

$$T = r c_o \dot{S}. \tag{3.10}$$

In addition, from (3.7), the temperature dependence of the sound propagation is made more explicit by assuming adiabatic compression and expansion of the waves. By selecting $c^2 = \mathrm{d}p/\mathrm{d}\rho$, (3.7) can be modified, but there are difficulties since c and u are in essence unknown functions of the position. To circumvent these difficulties, Lighthill expressed (3.7) in a slightly different form, which can be shown to be exact in general as follows [68]:

$$\sum_i \sum_j \frac{\partial^2}{\partial x_i \partial x_j}[(p - c_o^2 \rho)\delta_{ij}] = \sum_i \frac{\partial^2}{\partial x_i^2}(p - c_o^2 \rho) = \sum_i \frac{\partial^2 p}{\partial x_i^2} - c_o^2 \frac{\partial^2 \rho}{\partial x_i^2} \tag{3.11}$$

where δ_{ij} is the *Kronecker delta* with special values, $i = j : \delta = 1, i \neq j : \delta = 0$.
Thus from (3.5) and (3.6) becomes

$$\frac{\partial^2 \rho}{\partial t^2} - c_o^2 \sum_i \frac{\partial^2 \rho}{\partial x_i^2} = \frac{\partial \dot{Q}_m}{\partial t} + \sum_i \sum_j \frac{\partial^2 T_{ij}}{\partial x_i \partial x_j} - \sum_i \frac{\partial F_i}{\partial x_i} = \dot{S}, \, \mathrm{kgm}^{-3}\mathrm{s}^{-1}, \tag{3.12}$$

where

$$T_{ij} = \tau_{ij}^* + \rho u_i u_j + (p - c_o^2 \rho)\delta_{ij}, \text{ kgm}^{-1}\text{s}^{-2},\tag{3.13}$$

and c_o is the sonic speed for a gas at rest.

Thus the three terms on the right-hand side of (3.12) behave like the source terms discussed in Sect. 2.2, except that the source terms have dimension per unit volume in $\text{kgm}^{-1}\text{s}^{-2}$, instead of in $[\text{kgs}^{-2}]$. It is, therefore, necessary to examine carefully the question of whether a fluctuating-volume acoustical source may be considered equivalent to a point source. If the region of space occupied by the acoustic source has a small characteristic length scale λ, then this implies that the acoustic field observed in the far field is rather indifferent to the exact position of the source. We can, therefore, take the integral of the source, $\dot{S}$ $[\text{kgm}^{-1}\text{s}^{-2}]$ over the region $d\Omega$ $[\text{m}^3]$. Even though the source term contains terms due to volume force and turbulence, the units of all these terms are equivalent to the fluctuating mass flow rate, and we consider the effect of these all terms to be analogous to a fluctuating mass. Thus, in analogy to our discussion on fluctuating point mass source, point dipole source, and fluctuating quadrupole mass source, (3.12), we write the following expressions:

(a) Fluctuating mass (monopole):

$$\rho'(r,t) = \rho(r,t) - \rho_o = \frac{1}{c_o^2 4\pi r} \int \frac{\partial \dot{Q}_m}{\partial t}(0,t_1)d\Omega;\tag{3.14}$$

(b) Fluctuating volume force (dipole):

$$\rho'(r,t) = \frac{x_i}{c_o^2 4\pi r^2} \int \left[-\frac{1}{c_o}\dot{F}_i(0,t_1) - \frac{1}{r}F_i(0,t_1) \right] d\Omega;\tag{3.15}$$

(c) Fluctuating quadrupole:

$$\rho'(r,t) = \frac{x_i x_j}{c_o^2 4\pi r^3} \int \left[-\frac{1}{c_o^2}\ddot{T}_{ij}(0,t_1) - \frac{1}{c_o r}\dot{T}_i(0,t_1) + \frac{3}{r^2}T_{ij} \right] d\Omega.\tag{3.16}$$

Comparing (3.9) and (3.16) obviously

$$\rho' \simeq \frac{T_{ij}}{c_o^2 r^3}d\Omega.\tag{3.17}$$

It may be recalled from Chap. 2 that the highest time derivative term in (3.15), (3.16) remains only if $\varepsilon_2 = \lambda/(2\pi r) \ll 1$ (*far field condition*). If the condition is not satisfied, then the other terms need to be retained in the solution of the equations.

If the three terms on the right-hand side of (3.12) are zero, then the equation is reduced to a usual equation for the propagation of a sound wave in a medium of uniform temperature distribution:

$$\frac{\partial^2 \rho}{\partial t^2} - c_o^2 \sum_i \frac{\partial^2 \rho}{\partial x_i^2} = 0,\tag{3.18}$$

which can be solved along characteristic lines as an initial value problem. However, it was left to *Lighthill* to make an exact analogy of the nonzero terms on the right-hand side of (3.18), which will be discussed next.

3.2   Lighthill's Analogy

Mathematical tools to handle radiation problems are well developed in the field of electromagnetic and classical acoustic theories, in which a radiation source can be defined and separated from the radiation field it generates. Lighthill's formulation of aerodynamic noise theory is based on an acoustic analogy, which is an exact one.

Density fluctuation in a uniform acoustic medium at rest obeys the well-known inhomogeneous wave equation

$$\frac{\partial^2 \rho}{\partial t^2} - c_o^2 \sum_i \frac{\partial^2 \rho}{\partial x_i^2} = \dot{S}, [\mathrm{kg\,m^{-3}s^{-1}}]. \tag{3.19}$$

Comparing (3.19) with (3.12), there is an exact analogy as follows. The density fluctuation in a real fluid may be identified with fluids occurring in an ideal case of a uniform acoustic medium at rest produced by source terms. The general relation for density fluctuations under application of the foregoing analogy is

$$\rho(r,t) - \rho_o = \frac{1}{4\pi c_o^2} \int_\Omega \frac{\dot{S}}{r}(0, t - r/c_o)\mathrm{d}\Omega, [\mathrm{kg.m^{-3}}], \tag{3.20}$$

where Ω is the volume of the sound-producing zone.

Emphasized assumptions involved in writing the equation of density fluctuation are as follows:

(a) No back reaction from sound field to turbulence field is allowed.
(b) The process of generation of sound by a flow must be separated from the flow itself, and then consideration of far and near fields is necessary.
(c) The volumetric source terms must be considered restricted in a small region, that is, they must be considered as point sources.

Thus Lighthill's analogy allows simplification of the equations involved and estimation of the sound field in the following three steps:

Step 1: Compute the source field $\dot{S}$ from the given velocity or force distribution.
Step 2: Compute the acoustic noise.
Step 3: Make corrections for convection (to be discussed later).

In view of the fact that to validate Lighthill's analogy the observer must remain far from the sound-producing zone, the necessary conditions are described in Fig. 3.1.

$$\rho(x,t) - \rho_o = \frac{1}{4\pi c_o^2} \int \frac{\dot{S}}{|\mathbf{x} - \mathbf{y}|}\left(\mathbf{y}, t - \frac{|\mathbf{x} - \mathbf{y}|}{c_o}\right)\mathbf{y}\mathrm{d}\Omega(\mathbf{y}) \tag{3.21}$$

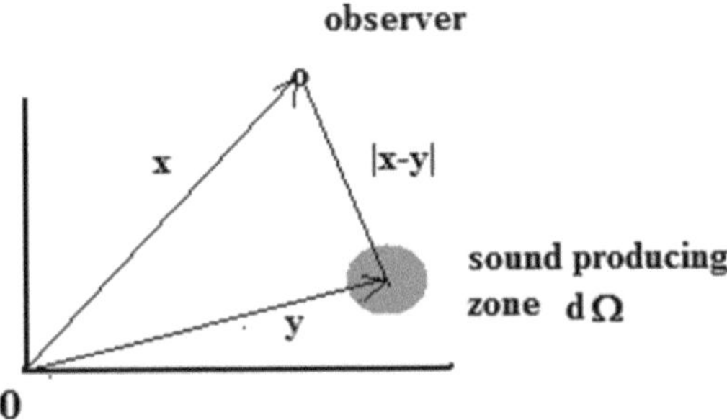

Fig. 3.1 Coordinates of a sound-producing zone and an observer

Consequently, for the three cases of fluctuating mass flow rate, fluctuating dipole, and fluctuating quadrupole, the corresponding equations are as follows (the dot gives the time derivative):

$$\rho(x,t) - \rho_o = \frac{1}{4\pi c_o^2} \int \frac{\dot{Q}_m}{|\mathbf{x}-\mathbf{y}|}\left(\mathbf{y},t - \frac{|\mathbf{x}-\mathbf{y}|}{c_o}\right) d\Omega(\mathbf{y}), \tag{3.22}$$

$$\rho(x,t) - \rho_o = \frac{1}{4\pi c_o^2} \int \left[-\frac{(x_i - y_i)}{c_o|\mathbf{x}-\mathbf{y}|} F_i\left(\mathbf{y},t - \frac{|\mathbf{x}-\mathbf{y}|}{c_o}\right) \right.$$
$$\left. -\frac{(x_i - y_i)}{|\mathbf{x}-\mathbf{y}|^3} F_i\left(\mathbf{y},t - \frac{|\mathbf{x}-\mathbf{y}|}{c_o}\right) \right] d\Omega(\mathbf{y}), \tag{3.23}$$

$$\rho(x,t) - \rho_o = \frac{1}{4\pi c_o^2} \int \left[-\frac{(x_i - y_i)(x_j - y_j)}{c_o|\mathbf{x}-\mathbf{y}|^3} \right.$$
$$\left. \left(\ddot{T}_{ij} + \frac{2c_o}{|\mathbf{x}-\mathbf{y}|}\dot{T}_{ij} + \frac{2c_o^2}{|\mathbf{x}-\mathbf{y}|^2}T_{ij}\right) \right] d\Omega(\mathbf{y}). \tag{3.24}$$

As discussed previously, only the highest derivative term will remain outside the primary sound-producing zone and in the far field. As a result of the analogy, the expressions for sound intensity, spectra, and acoustic power are derived.

From (1.108), the intensity of sound in (Wm^{-2}) at $\mathbf{x} = \mathbf{x}$ (x, y, z) is

$$I(\mathbf{x}) = \frac{c_o^3}{\rho}\overline{(\rho - \rho_o)^2}. \tag{3.25}$$

Now from (3.20) to (3.24), the general expression for the density fluctuation is

$$\rho(x,t) - \rho_o = \frac{1}{4\pi c_o^2} \int \varphi(\mathbf{x},\mathbf{y},c_o)\dot{S}\left(\mathbf{y},t - \frac{|\mathbf{x}-\mathbf{y}|}{c_o}\right) d\Omega, \tag{3.26}$$

where different expressions of φ in the far field for fluctuating mass flow, dipole, and quadrupole can be found by comparing (3.22) with (3.24).

Fig. 3.2 Two sound producing volumes as reaches the observer

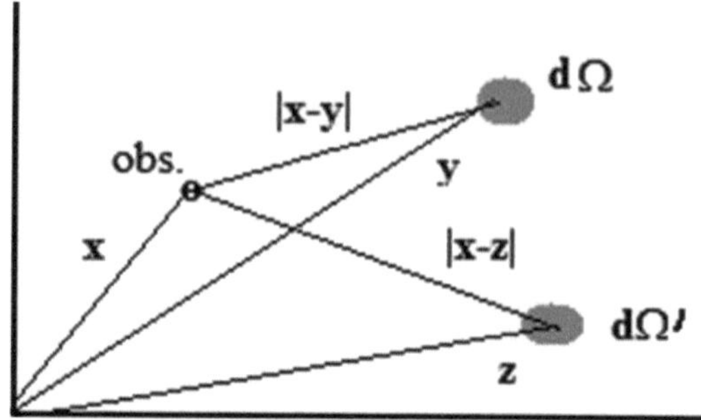

Fig. 3.3 Volume and correlation volume

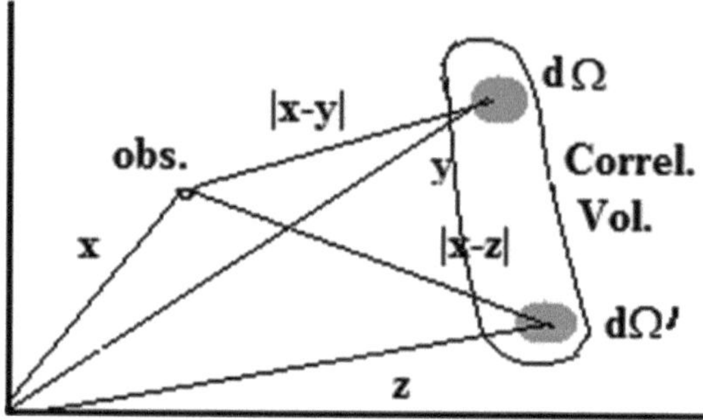

Writing (3.26) for two different places, $\mathbf{y}$ and $\mathbf{z}$ (Fig. 3.2), and with the help of (1.108), we obtain the relation for the intensity of sound as

$$I(\mathbf{x}) = \frac{1}{16\pi^2 c_o \rho_o} \int \int \varphi(\mathbf{x},\mathbf{y},c_o)\varphi'(\mathbf{x},\mathbf{z},c_o\dot{S})\left(\mathbf{y},t - \frac{|\mathbf{x}-\mathbf{y}|}{c_o}\right)\dot{S}'\left(\mathbf{z},t - \frac{|\mathbf{x}-\mathbf{z}|}{c_o}\right)\mathrm{d}\Omega(\mathbf{y})\mathrm{d}\Omega'(\mathbf{z}).$$

$$(3.27)$$

Equation (3.27) is very difficult to evaluate since there are difficulties in prescribing a volume and time arbitrarily. This is, therefore, done by prescribing a correlation volume (Fig. 3.3).

Now the following definitions are made:

$t_1 =$ time at which a sound wave originates at $\mathbf{y}$
$t_1' =$ time at which a sound wave originates at $\mathbf{z}$
$t_2 =$ time at which a sound wave from $\mathbf{y}$ reaches observer
$t_2' =$ time at which a sound wave from $\mathbf{y}$ reaches observer
$\Theta = t_1' - t_1 =$ time delay at source
$\Theta^* = t_2' - t_2 =$ time delay at observer

Let

$$B(\mathbf{x},\theta^*) = \overline{(\rho - \rho_o)^2} = \frac{1}{16\pi c_o^4}\int\int\varphi\varphi'\dot{S}(\mathbf{y},t_1)\dot{S}'(\mathbf{z},t_1 + \theta)\mathrm{d}\Omega(\mathbf{y})\mathrm{d}\Omega'.(\mathbf{z})\quad(3.28)$$

Since $d\Omega(\mathbf{y})$ and $d\Omega'(\mathbf{z})$ are rather arbitrary in nature, $\mathbf{z}$ is chosen such that $d\Omega'(\mathbf{z})$ is calculated as a *correlation volume*. Let $\delta = \mathbf{z} - \mathbf{y}$ and $d\Omega'(\mathbf{z}) = d\Omega^*(\delta)$, in which $d\Omega^*$ is the correlation volume around $\mathbf{y}$ and $d\Omega$ is the unit volume.

Noting the relationship between θ and θ^*,

$$\theta = \theta^* + \frac{|\mathbf{x} - \mathbf{y}| - |\mathbf{x} - \mathbf{z}|}{c_o}, \tag{3.29}$$

and from Fig. 3.2 it is shown especially for $|\delta| < |\mathbf{x}|$ that

$$|\mathbf{x} - \mathbf{y}| - |\mathbf{x} - \mathbf{z}| \approx \mathbf{CD} = \mathbf{BC}\cos\theta = \mathbf{BC}\frac{(\mathbf{x} - \mathbf{y})}{|\mathbf{x} - \mathbf{y}|} = \delta.\frac{(\mathbf{x} - \mathbf{y})}{|\mathbf{x} - \mathbf{y}|}, \tag{3.30}$$

and thus

$$\theta = \theta^* + \delta.\frac{(\mathbf{x} - \mathbf{y})}{|\mathbf{x} - \mathbf{y}|} \approx \theta^* + \delta.\frac{\mathbf{x}}{|\mathbf{x}|}. \tag{3.31}$$

It may further be noted that in Fig. 3.3 $\mathbf{y}$ and $\mathbf{z}$ are outside the reference point. However, a concept of correlation volume is introduced such that $\mathbf{y} = 0$. Thus the approximate relation in (3.31) is always valid if the correlation volume is taken. In addition, if δ and $\mathbf{x}$ are chosen perpendicular to each other, then $\theta = \theta^*$. Thus,

$$I(\mathbf{x}) = \frac{c_o^3 B(\theta^* = 0)}{\rho_o}, \tag{3.32}$$

which will be estimated in the next chapter.

3.3 Green's Formulation for Wave and Poisson Equation

The general wave equation in 3D space is given as

$$\frac{1}{c^2}\frac{\partial^2\rho'}{\partial t^2} - \nabla^2\rho' = 4\pi\dot{S}, \tag{3.33}$$

where the function $\dot{S}(\mathbf{r},t)$ describes the source density as being dependent on both time t and space $\mathbf{r}$. In addition, we need to specify the initial and boundary conditions. As is usual for such problems, where there is a distributed source, one can obtain the solution $G(\mathbf{r} - \mathbf{r}_o)$ at the observer point $\mathbf{r}$ caused by a unit source at the source point $\mathbf{r}_o$; then the field at $\mathbf{r}$ is caused by the source distribution and Gq is the integral over the whole range of the source at $\mathbf{r}_o$. The function G is called *Green's function*. It is written with two arguments, $\mathbf{r}$ and $\mathbf{r}_o$, separated by a vertical line.

To solve the general field problem, we seek the solution of the preceding equation for a point source of unit strength in the form of a *delta function* at $t = t_o$ and $\mathbf{r} = \mathbf{r}_o$, and we seek the solution of the equation with *Green's function* as

$$\frac{1}{c^2}\frac{\partial^2 G(r,t|\mathbf{r}_o,t_o)}{\partial t^2} - \nabla^2 G(\mathbf{r},t|\mathbf{r}_o,t_o) = 4\pi\delta(\mathbf{r}-\mathbf{r}_o).\delta(t-t_o) \qquad (3.34)$$

As an initial condition it may be considered reasonable to assume that G and $\partial G/\partial t$ are zero when $t < t_o$. In addition, we consider generally an outward propagating wave to $r \to \infty$, when G must vanish. Therefore, we seek a solution [81]

$$G(\mathbf{r}|\mathbf{r}_o) = \exp^{ikR}/R, \qquad (3.35)$$

where

$$k = (\omega/c)^2, \qquad (3.36)$$

$$R = |\mathbf{r}-\mathbf{r}_o| = \sqrt{(x-x_o)^2+(y-y_o)^2+(z-z_o)^2}. \qquad (3.37)$$

On the other hand, if we have a steady *Poisson equation* ($k = 0$) in the solution of the wave equation, then

$$\nabla^2\rho = -4\pi\dot{S}(\mathbf{r},t) \qquad (3.38)$$

and *Green's function* in three dimensions can be written as $G(\mathbf{r}—\mathbf{r}_o) = 1/R$.

3.4 Exercises

3.4.1 What are the main points of *Lighthill's analogy*?

3.4.2 What is the physical meaning of the term *correlation volume*?

3.4.3 Find the value quadrupole moment from the expression $T_{ij} \simeq \rho u_i' u_j' \simeq \rho_o u_i'^2 [\mathrm{Jm}^{-3}]$ for air density $\rho_o = 1.26[\mathrm{kgm}^{-3}]$ and the subsonic fluid velocity $U \approx 200[\mathrm{ms}^{-1}]$, assuming approximately 2% turbulence energy.

3.4.4 Taking the value of the quadrupole moment of the previous problem, and taking $c_o = 330.0\,\mathrm{ms}^{-1}$, $r = 20\,\mathrm{m}$, and $1\,\mathrm{cm}^{-3}$ as the fluctuating volume, compute the average perturbed density $[\mathrm{kg\,m}^{-3}]$.

Chapter 4
Subsonic Jet Without Considering Convection

4.1 Dimensional Analysis by Lighthill

For subsonic jets, as shown schematically in Fig. 4.1 for a 2-dimensional or axisymmetric jet emanating parallel to the axis, different regions can be described. It is found that for moderate gas velocities and on the axis of the jet, the velocity remains constant up to $x/D_o = 4$, in which x is the coordinate beginning from the exit plane and D_o is the exit jet height or the exit diameter. The velocity on the axis of the jet decreases rapidly beyond $x/D_o > 4$, and there are similar velocity profiles beyond $x/D_o > 6$, whereas y/D_o increases continuously. In addition, from another experiment with three different gases, Lassiter and Hubbard [56] found that larger jets generate noise in which the low-frequency component is somewhat higher than in smaller jets. Furthermore, a higher frequency emanates from a point just outside the jet, whereas the low-frequency components come from $x/D_o = 3$ to 5.

In his first treatise on aerodynamic noise, Lighthill did not take into account the variation of the frequency in the jet, which was later included by Ribner [96] and Powell [90].

The equation for the production of the turbulent jet noise is written from (3.28) in which, for the source term, the appropriate quadrupole term is replaced by (3.24). Thus, one gets

$$B(x, \theta^* = 0) = \overline{(\rho - \rho_o)^2}$$

$$= \frac{1}{16\pi^2 c_o^4} \int \int \frac{(x_i - y_i)(x_j - y_j)(x_k - y_k)(x_l - y_l)}{c_o^4 |\mathbf{x} - \mathbf{y}|^4}$$

$$\overline{\ddot{T}_{ij}(\mathbf{y}, t) \ddot{T}_{kl}(\mathbf{y} + \delta, t + \theta)} \, \mathrm{d}\Omega (\mathbf{y} \mathrm{d}\Omega^*(\delta), \tag{4.1}$$

and the intensity of sound radiation is

$$I(\mathbf{x}) = \frac{c_o^3 B(\mathbf{x}, \theta^* = 0)}{\rho_o}. \tag{4.2}$$

T. Bose, *Aerodynamic Noise: An Introduction for Physicists and Engineers*,
Springer Aerospace Technology 7, DOI 10.1007/978-1-4614-5019-1_4,
© Springer Science+Business Media New York 2013

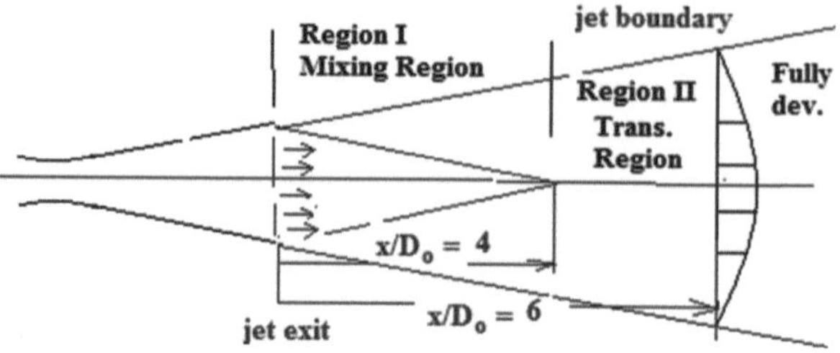

Fig. 4.1 Different regions of a jet. (I) Mixing region, and (II) Transition region followed by fully-developed region

For a constant radian frequency, ω, there will be a phase shift between fluctuations emanating at two different places if measurements are taken simultaneously, and because there is an additional time difference θ, the following fluctuations for T_{ij} and T_{kl} are assumed:

$$T_{ij} = A_1(\mathbf{y})\exp^{j\omega t} \text{ and } T_{kl} = A_2(\delta)\exp^{j\omega_\delta \delta}\exp^{j\omega(t+\theta)}. \tag{4.3}$$

Thus, it can be shown that

$$\frac{\partial^4 T_{ijkl}}{\partial\theta^4}(\mathbf{y},\delta,\theta) = \overline{\frac{\partial^2 T_{ij}}{\partial t^2}\frac{\partial^2 T_{kl}}{\partial t^2}}, \tag{4.4}$$

where

$$T_{ijkl} = \overline{T_{ij}(\mathbf{y},t)T_{kl}(\delta,t+\theta)} \tag{4.5}$$

and θ is computed from (3.31) for $\theta^* = 0$, that is, $\theta = \delta \cdot \mathbf{x}/|\mathbf{x}|$. For the quadrupole moment T_{ij} is approximated from (3.13) by the relation $T_{ij} \simeq \rho_o u_i u_j$. With $|\mathbf{x} - \mathbf{y}| = r$, $L^3 \sim d\Omega^*$, where L is a characteristic dimension of the correlation volume Ω^*, the velocities are proportional to a characteristic velocity U, and time θ is inversely proportional to the characteristic frequency of the acoustic radiation; we can write from (4.1) and (4.2) that

$$dB \sim \frac{\rho_o^2 U^4 v^4 L^3}{c_o^8 r^2}d\Omega \text{ and } dI \sim \frac{c_o^3}{\rho_o}dB \sim \frac{\rho_o U^4 v^4 L^3}{c_o^5 r^2}d\Omega \tag{4.6}$$

and the acoustic power

$$dP = 4\pi r^2 dI(r) \sim r^2 dI \approx \frac{\rho_o U^4 v^4 L^3}{c_o^5}d\Omega. \tag{4.7}$$

For the characteristic length L and for the volume $d\Omega$, Lighthill [63], on the one hand, and Ribner [96] and Powell [90], on the other, give slightly different estimates and, thus, slightly different results. Ribner and Powell [90] divide the subsonic jet region into regions I and II, as shown in Fig. 41, in which region I extends within the *mixing region* and region II within the transition and fully developed region.

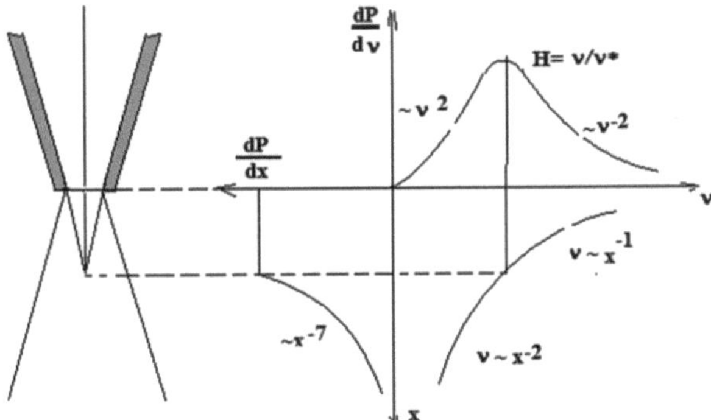

Fig. 4.2 Acoustic power and frequency distribution in a jet

The estimate for the acoustic power spectra is done by assuming that a given slice of jet emits just a single frequency v. Thus the spectrum emitted by the jet as a whole can be approximated as the *acoustic power spectrum*:

$$\frac{\mathrm{d}P}{\mathrm{d}v} = \frac{\mathrm{d}P}{\mathrm{d}x} \cdot \frac{\mathrm{d}x}{\mathrm{d}v} \tag{4.8}$$

By slight modification of the method of *Lighthill*, *Ribner* and *Powell* give, therefore, a new relation for acoustic power:

$$\frac{\mathrm{d}P}{\mathrm{d}v} = \frac{\rho_o U_{\mathrm{jet}}^7 D_o^3}{c_o^5} H\left(\frac{v}{v^*}\right), \quad v^* = \frac{U_{\mathrm{jet}}}{D_o}, \tag{4.9}$$

where $H(v/v^*)$ is the bell-shaped spectrum. *Ribner* proposed a method to show his results on the basis of dimensionless analysis, which is given schematically in Fig. 4.2. It can be seen that the *acoustic power* distribution per unit length remains fairly constant in region I and decreases with the seventh power of x in region II. These results seem to agree with the experimental observations.

Estimates of the sound intensity, *acoustic power*, and *acoustic power spectra* are given in Table 4.1 separately for Lighthill [63] and Ribner [96] and Powell [90]. Thus we get the celebrated U_{jet}^8 law by *Lighthill* for subsonic jets. *Lighthill* himself suggested that for the same thrust but using a larger jet exit area it should be possible to reduce the noise level. This prediction was utilized in the design of fanjet engines, in which the low average jet speed not only reduced noise but also at the same time the *propulsive efficiency* of the jet increased. Reduction of the noise is also possible by reducing the jet exit diameter (by partitioning the total flow area!) and by reducing the shear in the mixing region; this has created fascinating opportunities for the design of multiple nozzles and multiple lobe nozzles (Fig. 4.3).

On the basis of dimensionless analysis, Lighthill proposed a coefficient of proportionality of the *acoustic power*

$$K = \frac{\text{acoustic power}}{\rho_o D_o^2 c_o^{-5}} \tag{4.10}$$

and suggested its measurement. Subsequently, for subsonic jets its value was found to be $K = 10^{-4}$.

Table 4.1 Dimensional analysis of acoustic quantities

Description	Lighthill	Ribner and Powell	
		Region I	Region II
1. Scale length L	$L \sim D_o$	$L \sim x$	$L \sim x$
2. Velocity U	$U \sim U_{jet}$	$U \sim U_{jet}$	$U \sim U_{jet}D_o/x$
3. Volume Ω or $d\Omega$	$d\Omega \sim A\,d\,x \sim D_o x\,dx$	$d\Omega \sim Adx \sim x^2 dx$	
4. Frequency	$v \sim U_{jet}/D_o$	$v \sim U/x \sim U_{jet}/x$	$v \sim U/x \sim U_{jet}D_o/x^2$
5. Sound intensity I or dI	$I \sim \frac{\rho_o U_{jet}^8}{c_o^5}\left(\frac{D_o}{r}\right)^2$	$dI \sim \frac{\rho_o U_{jet}}{8D_o}c_o^5 r^2 dx$	$dI \sim \frac{\rho_o U_{jet}}{8D_o^8}c_o^5 x^7 r^2 dx$
6. Acoustic power P or dP	$P \sim \frac{\rho_o U_{jet}^8}{c_o^5}$	$dP \sim \frac{\rho_o U_{jet}}{8D_o}c_o^5 dx$	$dP \sim \frac{\rho_o U_{jet}}{8D_o^8}c_o^5 x^7 dx$
7. Acoustic power per length $\frac{dP}{dx}$ $\frac{dP}{dx} \sim \frac{\rho_o U_{jet}}{8D_o^8}c_o^5 x^7$	–	$\frac{dP}{dx} \sim \frac{\rho_o U_{jet}}{8D_o}c_o^5$	$\frac{dP}{dx} \sim \frac{\rho_o U_{jet}}{8D_o}c_o^5$
8. $\frac{dv}{dx}$	–	$\frac{dv}{dx} \sim v^2/U_{jet}$	$\frac{dv}{dx} \sim \frac{U_{jet}D_o}{x^3}$
9. Acoustic power spectra	$\frac{P}{v} \sim \frac{\rho_o U_{jet}^7{}^8}{c_o^5}$	$\frac{dP}{dv} \sim \frac{\rho_o U_{jet}^7 D_o^8 x}{c_o^5}dx \sim \frac{\rho_o U_{jet}^9 D_o v^2}{c_o^5}$	$\frac{dP}{dv} \sim \frac{\rho_o U_{jet}^7 D_o^8 x}{c_o^5 x^2}dx \sim \frac{\rho_o U_{jet}^5 D_o v^2}{c_o^5}$

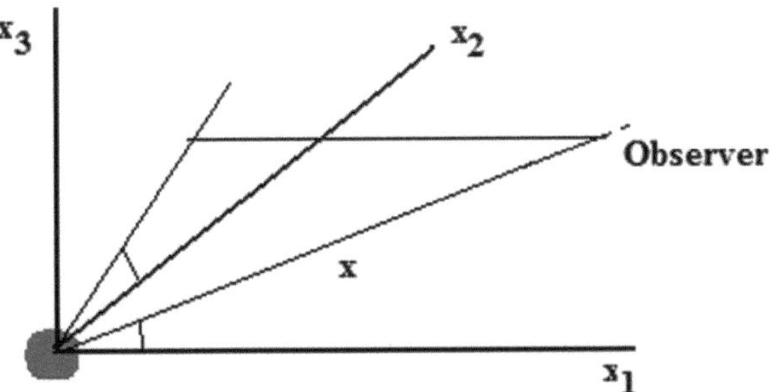

Fig. 4.3 Coordinates for a distant observer to compute acoustic power

Table 4.2 Acoustic power for various acoustic sources

Source	Acoustic power (W)
Voice – very soft whisper	10^{-9}
Voice – conversational level	10^{-5}
Voice – shouting, ventilator fan	10^{-3}
Car on highway, 1.3 m loom	10^{-2}
Radio	10^{-1}
Piano, large chipping hammer	10^{0}
75-piece orchestra, small aircraft engine	10^{1}
4-propeller airliner	500
Turbojet (3,500 kg_f thrust)	10^{4}
Ramjet, turbojet with afterburner	10^{5}

Further, *Lighthill* proposed an efficiency of sound production, defined as the ratio of the *acoustic power* to the development of the power in a jet,

$$\eta = \frac{\text{acoustic power}}{\text{supply of power}} = \frac{K\rho_o U_{\text{jet}}^8 c_o^{-5} D_o^2}{\rho_o U_{\text{jet}}^3 D_o^2} \, M_{\text{jet}}^5. \tag{4.11}$$

As a guideline the acoustic power for various acoustic sources are given in Table 4.2.

In a recent paper by Garrison et al. [41], a "Computational Fluid Dynamics (CFD) analysis using the *Reynolds averaged Navier–Stokes equations (RANS)* with the two-equation Shear Stress Transport (SST) turbulence model is performed for internally mixed jets with the goal of accurately predicting the development of the turbulence in the resulting jet plumes." These forced mixtures are put in multilobe form for the core flow inside a bypass flow before the exit nozzle for the low-bypass turbojet engine and shows a reduction by several decibels in various frequencies.

4.2   Self and Shear Noise

Lighthill's treatise was not in a position to describe the directivity of subsonic jet noise. This was done by Csanady [34] and Ribner [95, 96] by examining in detail the quadrupole noise term by *Lighthill*, in which case it was necessary to examine the combination possibilities for the *autocorrelation* of quadrupole moments in different directions.

The starting equation for an investigation of these possibilities is the equation for the intensity of sound (3.27), in which it is assumed that $\mathbf{x} \gg \mathbf{y}$:

$$I(\mathbf{x}) = \frac{1}{16\pi^2 c_o \rho_o} \frac{x_i x_j x_k x_l}{x^6} \int \int \ddot{T}_{ij} \ddot{T}_{kl} \, d\Omega(\mathbf{y}) d\Omega'(\delta). \qquad (4.12)$$

The acoustic power radiated from the sound-producing zone is obtained by integrating the intensity through a spherical surface A_S around the sound-producing zone

$$P = \int \int I \, dA_S. \qquad (4.13)$$

Evaluations of (4.5) and (4.13) are done in the spherical coordinates, for which the following relations may be noted:

$$x_1 = x \cos\theta,$$

$$x_2 = x \sin\theta \cos\varphi,$$

$$x_3 = x \sin\theta \sin\varphi,$$

$$dA_S = x^2 \sin\theta \, d\varphi d\theta.$$

Combining (4.5) and (4.13) we can now write

$$P = \frac{1}{16\pi^2 c_o \rho_o} \int \int \int \int \frac{x_i x_j x_k x_l}{x^4} \ddot{T}_{ij} \ddot{T}_{kl} \, d\Omega(\mathbf{y}) d\Omega'(\delta) \sin\theta \, d\varphi d\theta. \qquad (4.14)$$

By defining a *correlation integral*

$$C_{ijkl} = \frac{1}{16\pi^2 c_o \rho_o} \int \int \ddot{T}_{ij} \ddot{T}_{kl} \, d\Omega(\mathbf{y}) d\Omega'(\delta) \qquad (4.15)$$

we can rewrite (4.14) as

$$P = \int_{\varphi=0}^{2\pi} \int_{\theta=0}^{\pi} \frac{x_i x_j x_k x_l}{x^4} C_{ijkl} \sin\theta \, d\varphi d\theta, \qquad (4.16)$$

in which the integration must be performed for φ between 0 and 2π, and for θ from 0 to π. To have an expression for the *directivity pattern*, we introduce further $P^* = P/C_{ijkl}$, and we get

$$P^* = \int_{\varphi=0}^{2\pi} \int_{\theta=0}^{\pi} \frac{x_i x_j x_k x_l}{x^4} \sin\theta \, d\varphi d\theta. \qquad (4.17)$$

If the x_1-axis coincides with the jet axis, then a differentiation of (4.17) with respect to θ gives the *directivity pattern* of the jet noise. The analysis is, however, much simplified if proper valid combinations of subscripts i, j, k, and l are considered. For this purpose, it will be good to keep the following integrals in mind:

$$P^* = \int_{\varphi=0}^{2\pi} \sin^4 \varphi \, d\varphi = \int_{\varphi=0}^{2\pi} \cos^4 \varphi \, d\varphi = \frac{3\pi}{4},$$

$$P^* = \int_{\theta=0}^{\pi} \sin^5 \theta \, d\theta = \int_{\theta=0}^{\pi} \cos^5 \theta \, d\theta,$$

$$P^* = \int_{\theta=0}^{\pi} \cos^4 \theta \sin \theta \, d\theta = \frac{2}{5}.$$

Now the different combinations of i, j, k, and l are evaluated as follows:

1. $i = j = k = l$. By taking $i = 1$, 2, or 3, (4.17) is integrated to get

$$P^* = \int_{\varphi=0}^{2\pi} \int_{\theta=0}^{\pi} \left(\frac{x_i}{x}\right)^4 \sin \theta \, d\varphi \, d\varphi = \frac{4\pi}{5}. \tag{4.18}$$

2. i and j are in pairs, that is, $i = j$ and $k = l$, or $i = k$ and $j = l$, etc., and we get

$$P^* = \int_{\varphi=0}^{2\pi} \int_{\theta=0}^{\pi} \left(\frac{x_i^2 x_j^2}{x^4}\right) \sin \theta \, d\varphi \, d\varphi = \frac{4\pi}{15}. \tag{4.19}$$

3. There is only one pair, for example, 1123 or 1112:

$$P^* = \int_{\varphi=0}^{2\pi} \int_{\theta=0}^{\pi} \left(\frac{x_1^2 x_2 x_3}{x^4}\right) \sin \theta \, d\varphi \, d\varphi = \frac{4\pi}{15}$$

$$= \int \int \cos^2 \theta \sin^3 \theta \cos \varphi \sin \varphi \, d\theta \, d\varphi.$$

Since

$$\int_{\varphi=0}^{2\pi} \cos \varphi \sin \varphi \, d\varphi = \frac{1}{2} \int_{\varphi=0}^{2\pi} \sin 2\varphi \, d\varphi = 0, \tag{4.20}$$

obviously $P_{1123}^* = 0$.

Thus from the total number of 81 possible combinations of indices for the correlation integrals, only 9 have nonzero values, and these are as follows:

(i) *Longitudinal correlations*: $i = j = k = 1$, and thus there may be three combinations of 1111, 2222, or 3333, that is, there are three values.

(ii) *Lateral correlations*: i and j in pairs, that is, the possible combinations are 1212, 1313, 2323, 1122, 1133, and 2233, that is, six values.

Not counted are the combinations 2121, 3131, 3232, 2211, 3311, 3322, 1221, 1331, 2332, 3223, and 2112, which are mere repetitions.

Now we examine the possible combinational correlations of turbulent fluctuations, for which we take the x_1-axis along the axis of the jet and the x_2-axis in the radial direction. Since the important velocity gradient for the flow will be $\partial u_1/\partial x_2 \neq 0$, we examine the effect of the velocity gradient in the production of sound.

From (3.13), a reasonable estimate for the quadrupole term is $T_{ij} = \rho u_i u_j$. Let

$$\rho = \bar{\rho} + \rho';\, u_i = \bar{u}_i + u'_i;\, u_j = \bar{u}_j + u'_j. \tag{4.21}$$

Further, let

$$\frac{\rho}{\bar{\rho}} = 1 + \frac{\rho'}{\bar{\rho}} = \left(\frac{T}{\bar{T}}\right)^{1/(\gamma-1)} = \left(1 + \frac{T'}{\bar{T}}\right)^{1/(\gamma-1)}$$

$$c_p(T - \bar{T}) = c_p T' = (V^2 - \bar{V}^2)/2 \approx \bar{V}V'. \tag{4.22}$$

Hence

$$\frac{T}{\bar{T}} = 1 + (\gamma - 1)\bar{M}^2\left(\frac{V'}{\bar{V}}\right), \tag{4.23}$$

$$T' = \frac{VV'}{2c_p} = \frac{(\gamma - 1)\bar{M}^2\bar{T}}{2V'^2/\bar{V}}, \tag{4.24}$$

$$\frac{\rho'}{\bar{\rho}} = \left[1 + \frac{\gamma - 1}{2}M^2\frac{V'}{V}\right]^{\frac{1}{\gamma-1}} - 1 \approx \frac{1}{2}M^2\frac{V'}{V}. \tag{4.25}$$

Thus, as long as $\bar{M} \ll 1$, $\rho'/\bar{\rho}$ can be neglected. Therefore, under these conditions only,

$$T_{ij} = \bar{\rho}\bar{u}_i\bar{u}_j + \bar{\rho}\bar{u}_i u'_j + \bar{\rho}u'_j\bar{u}_i + \bar{\rho}u'_i u'_j.$$

On the right-hand side of the preceding equation, the first term does not contribute to the production of sound, and the second and third terms will contribute if there is a shear layer, for example, $\partial u_1/\partial x_2 \neq 0$. Similarly, from (3.18) and (3.19) for an axisymmetric case, the term for the quadrupole fluctuations is $\partial^2(T_{ij}/r)/(\partial x_i \partial x_j)$. By setting $i = 1$ and $j = 2$, we get

$$\frac{\partial^2}{\partial x_1 \partial x_2}\left(\frac{T_{12}}{r}\right) = \bar{\rho}\frac{\partial^2}{\partial x_1 \partial x_2}\left(\frac{\bar{u}_1 u'_2}{r}\right) = \bar{\rho}\left[\frac{x_1 - y_1}{r^2}\frac{\partial \bar{u}_1}{\partial x_2}\left(\frac{\partial u'_2}{\partial r} - \frac{u'_2}{r}\right)\right.$$

$$\left. + \frac{(x_1 - y_1)(x_2 - y_2)}{r}\left(\frac{1}{r^2}\frac{\partial^2 u'_2}{\partial r^2} - \frac{2}{r^2}\frac{\partial u'_2}{\partial r} + \frac{3u'_2}{r^4} - \frac{1}{r^3}\frac{\partial u'_2}{\partial r}\right)\right].$$

With $\partial/\partial r = -c_o^{-1}\partial/\partial t$ and taking the highest terms in the far field only, we get

$$\frac{\partial^2}{\partial x_1 \partial x_2}\left(\frac{T_{12}}{r}\right) = \bar{\rho}\left[-\frac{x_1 - y_1}{r^2 c_o}\frac{\partial \bar{u}_1}{\partial x_2}\ddot{u}_2 + \frac{(x_1 - y_1)(x_2 - y_2)}{r}\frac{\ddot{u}_2}{c_o^2 r^2}\right]. \tag{4.26}$$

While the second term is easily identified as of a quadrupole nature, the first term is a dipole. The sound generated by the first term is called the *self noise* and the second term is known as the *self noise*.

Ribner has shown how these can be estimated. The acoustic energy emanating from the unit volume in the direction of the observer and per unit of the solid angle is given by the relation at large distances as

$$\frac{\partial^2 P(\theta,\varphi)}{\partial\theta\partial\varphi} = \frac{x_i x_j x_k x_l}{16\pi^2 \rho_o c_o^5 x^4}\int\sqrt{\frac{\partial^2 T_{ij}}{\partial r^2}\frac{\partial^2 T_{kl}}{\partial r^2}}\,\mathrm{d}\Omega^*$$

$$= \frac{x_i x_j x_k x_l}{16\pi^2 c_o^5 x^4}\int\frac{\partial^4}{\partial\tau^4}(u_i u_j u_k u_l)\mathrm{d}\Omega^*. \tag{4.27}$$

The relative *angular directivity patterns* for the different quadrupoles for the *self* and *shear* noises are as follows:

(a) Self noise:

Longitudinal quadrupole self-correlation: $T_{11}T_{11} \sim \cos^4\theta$,

Lateral quadrupole self-correlation: $T_{12}T_{12} \sim \sin^2\theta\cos^2\theta$,

Lateral quadrupole self-correlation: $T_{13}T_{13} \sim \sin^2\theta\cos^2\theta$,

Lateral quadrupole cross correlation: $T_{11}T_{33} \sim \sin^2\theta\cos^2\theta$,

Longitudinal quadrupole self-correlation: $T_{22}T_{22} \sim \sin^4\theta$,

Lateral quadrupole self-correlation: $T_{23}T_{23} \sim \sin^4\theta$,

Lateral quadrupole self-correlation: $T_{22}T_{33} \sim \sin^4\theta$.

Adding a contribution from all these and noting their relative values, *Ribner* showed that the self noise must be independent of θ.

(b) Shear noise:

Longitudinal quadrupole self-correlation: $T_{11}T_{11} \sim \cos^4\theta$,

Lateral quadrupole self-correlation: $T_{13}T_{13} \sim \sin^2\theta\cos^2\theta$.

Adding a contribution from all quadrupoles, the intensity of sound from a round jet is given by

$$I \sim A + B\frac{(\cos^4\theta + \cos^2\theta)}{2}. \tag{4.28}$$

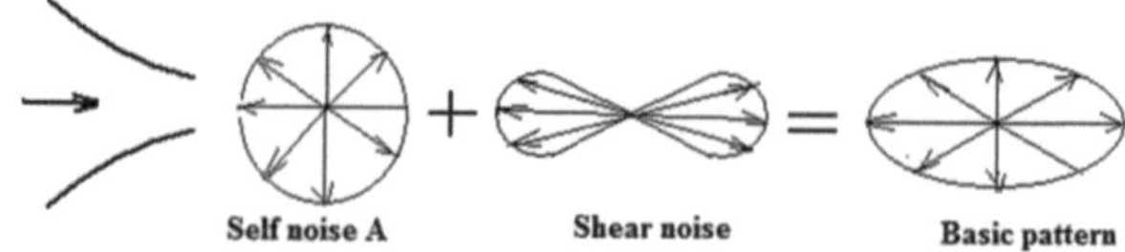

Fig. 4.4 Basic subsonic jet noise patten

The nondimensional self noise results from the joint contribution of nine quadrupole correlations having $\cos^2\theta\sin^2\theta$ or $\sin 4\theta$ directionality, whereas the second term comes from the shear noise. These are shown schematically in Fig. 4.4.

4.3 Estimation of Jet Noise by Proudman

Proudman's estimation of jet noise was the first application of Lighthill's jet noise theory, and he shows how, due to friction, the turbulent energy is decayed and converted into acoustic energy. The assumptions made in his analysis are that the turbulence is homogeneous and isotropic, and the sound intensity and power are measured in the direction of the correlation distance vector, $\delta \cdot \mathbf{x} = 0$. Thus only the longitudinal quadrupole need be considered, which means that $T_{ij} \approx \rho_o \overline{u'^2}$. From (4.1) and (4.2) we get

$$\frac{\mathrm{d}P}{\mathrm{d}\Omega} = \frac{\rho_o}{4\pi c_o^5} \int \overline{\frac{\partial^2}{\partial t^2} u'^2(0) u'^2(\delta)} \,\mathrm{d}\Omega. \tag{4.29}$$

Now for this type of correlation function for u' having some experimental support, we may write for the integral

$$\int \overline{\frac{\partial^2}{\partial t^2} u'^2(0,t) u'^2(\delta,t)}\,\mathrm{d}\Omega = -152\pi \left(\overline{u'^2}\right)^{5/2} \frac{\mathrm{d}E_{\mathrm{turb}}}{\mathrm{d}t}, \tag{4.30}$$

where the *turbulent kinetic energy* is given by

$$\mathrm{d}E_{\mathrm{turb}} = \frac{3}{2}\frac{\overline{u'^2}}{2} \tag{4.31}$$

and $\mathrm{d}E_{\mathrm{turb}}/\mathrm{d}t$ gives the *mean rate of dissipation of turbulent energy*. Thus we get the relation for the acoustic power as

$$\frac{\mathrm{d}P}{\mathrm{d}\Omega} = -38\frac{\rho_o}{c_o^5}\left(\overline{u'^2}\right)^{5/2}\frac{\mathrm{d}E_{\mathrm{turb}}}{\mathrm{d}t}. \tag{4.32}$$

From the theory of *homogeneous turbulence*,

$$\frac{dE_{\text{turb}}}{dt} = -30\frac{\mu}{\rho}\frac{\overline{u'^2}}{L_f^2},$$
(4.33)

where μ is the dynamic viscosity coefficient and L_f is the *scale of the longitudinal turbulence*. Therefore, for the volumetric acoustic power we get

$$\frac{dP}{d\Omega} = 1140\frac{\mu}{c_o^5}\left(\frac{\rho_o}{\rho}\right)\left(\frac{\overline{u'^2}}{L_f^2}\right)^{7/2}.$$
(4.34)

Equations (4.33) and (4.34) show that for small-scale turbulence,

$$L_f \to 0, \frac{dP}{d\Omega} \to \text{large and } \left|\frac{dE_{\text{turb}}}{dt}\right| \to \text{large}$$

and the turbulent kinetic energy is dissipated fastest and generates maximum sound. In the mixing region of a subsonic jet L_f is small, which explains why the maximum acoustic power is delivered from the mixing zone of a subsonic jet. Further, since from dimensional analysis the frequency can be obtained by dividing the jet velocity U_{jet} by the characteristic length v U_{jet}/L_f, a natural corollary to the Proudmann estimate of sound is that high-frequency turbulence is more effective for sound generation.

4.4 Exercises

4.4.1 Explain, with the help of a small sketch, the different regions of a subsonic jet with an exit diameter of 4 cm.

4.4.2 What are the *frequency* and *power spectrum* distributions in a subsonic jet?

4.4.3 What is the *acoustic power* of the various sources?

4.4.4 What are *self* and *shear noises*? What is the *basic pattern* of a subsonic jet noise, combining the two?

Chapter 5
Subsonic Jet Noise (Including Effect of Convection)

In Chap. 4, the equations for the intensity of sound and acoustic power were developed as if, due to fluctuations in the flow properties, there were a fluctuation in the quadrupole strength producing noise, but otherwise the quadrupole remained stationary in space. In this chapter, these are now modified to account for the relative velocity between the source and the observer.

Motion of air or gas has two different effects on the propagation of sound. When the source and the observer (receiver) are moving relative to each other, there is a shift in the frequency and the wavelength of the radiating sound, which is called the *Doppler effect* and will be discussed in the following section. This is also found in the case of propagation of electromagnetic waves and is dealt with by the transformation of the coordinate by the *Lorentz transformation*. However, the transformation is based on the idea that there is an absolute limitation in the velocities of many bodies imposed by the velocity of propagation of electromagnetic waves. An important difference between the propagation of sound waves and the propagation of electromagnetic waves is that the velocity of energy flow for sound waves at any point is the vector sum of the velocity of sound in a quiescent atmosphere and the velocity of the propagating medium, whereas such an effect is not known for electromagnetic waves. In the presence of a wind gradient, this effects the line of propagation of sound waves such that they bend in the downstream direction (*refraction of sound waves*), and the sound level at the receiver is much greater when it is in a downward position than in an upwind position. This leads to a *shadow region* in the upwind direction, depending on the frequency of the propagating sound wave and the local velocity of the sound wave (local temperature). There is a strong reduction in intensity of the upwind direction, but it is amplified in the downwind direction. For a detailed discussion of the effect of wind and temperature gradients on the propagation of sound waves, readers are referred to Harris [4] and references therein.

T. Bose, *Aerodynamic Noise: An Introduction for Physicists and Engineers*,
Springer Aerospace Technology 7, DOI 10.1007/978-1-4614-5019-1_5,
© Springer Science+Business Media New York 2013

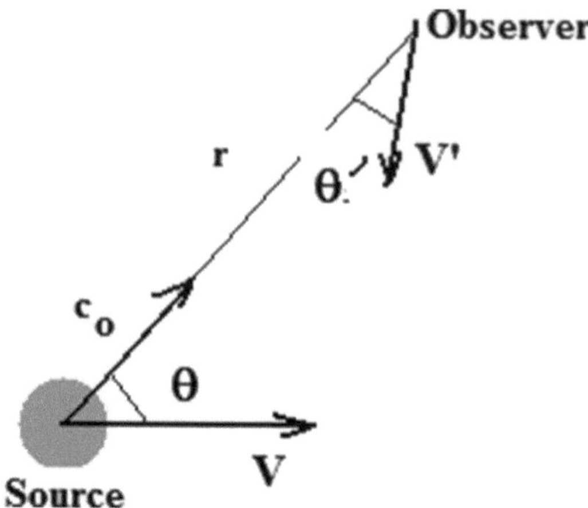

Fig. 5.1 Moving source/observer

5.1 Doppler Effect

If an observer and a source of sound are not moving with respect to each other (the distance between them being r), but the latter sends sound waves with a wavelength λ_o, then the number of oscillations reaching the observer in a time unit is given by the frequency

$$v_o = c_o/\lambda_o, \tag{5.1}$$

where c_o is the velocity of propagation of the sound waves (in ambient air at rest).

If only the observer is moving toward the source, then in time t the observer travels a distance of $[c_\infty + (V'\cos\theta)]t = ct$, and then

$$v = \frac{c_o + (V'\cos\theta')}{\lambda_o} = \frac{c_o}{\lambda_o}\left[1 + \frac{V'}{c_o}\cos\theta'\right] = v_o(1 + M'\cos\theta'), \tag{5.2}$$

On the other hand, if the source is moving, then in the time period $T = 1/v_o$ the observer covers a distance $l = VT$, and there is effectively a compression of the wavelength to the effective wavelength

$$\lambda = \lambda_o - l\cos\theta = \lambda_o\left(1 - \frac{l}{\lambda_o}\cos\theta\right)$$

$$= \lambda_o(1 - M\cos\theta). \tag{5.3}$$

Therefore, the observer accepts a corresponding frequency of

$$v = \frac{c_o}{\lambda} = \frac{c_o}{\lambda_o(1 - M\cos\theta)} = \frac{v_o}{1 - M\cos\theta}. \tag{5.4}$$

Table 5.1 Regarding Doppler effect

Wave properties	Observer moving alone	Source moving alone	Both moving
Wavelength	$\lambda = \frac{\lambda_o}{1+M'\cos\theta'}$	$\lambda = \lambda_o(1-M\cos\theta)$	$\lambda = \lambda_o\frac{(1-M\cos\theta)}{(1+M'\cos\theta')}$
Frequency	$v = v_o(1+M'\cos\theta')$	$v = \frac{v_o}{(1-M\cos\theta)}$	$v = v_o\frac{(1+M'\cos\theta')}{(1-M\cos\theta)}$

These results are summarized in Table 5.1, which shows an apparent discrepancy in equations for an observer moving alone and a source moving alone since from the relativity principle they must appear to be the same.

However, from the principles of relativity, the compression or the expansion factor is $\sqrt{[1+(M\cos\theta)^2]}$ instead of $(1+M\cos\theta)$, and it can be shown by binomial expansion that for $M\cos\theta \ll 1$ both expressions are the same. However, there seems to be no experimental verification of the discrepancies at different Mach numbers.

The number of waves n that can be counted to cover a distance in a certain time period because the number of waves reaching the observer, whether stationary or moving, is the same is given by the relation

$$t = \frac{n\lambda}{c_o} = n\frac{n\lambda_o}{c_o}\frac{(1-M\cos\theta)}{(1+M'\cos\theta')} = l_o\frac{(1-M\cos\theta)}{(1+M'\cos\theta')}, \tag{5.5}$$

and the apparent distance is

$$r = c_o t = n\lambda = n\lambda_o\frac{(1-M\cos\theta)}{(1+M'\cos\theta')} = r_o\frac{(1-M\cos\theta)}{(1+M'\cos\theta')}. \tag{5.6}$$

Thus the wave will be shifted toward the shorter wavelength if the source is moving toward the observer and to longer wavelength if the source is moving away from the observer.

5.2 Experimental Determination of the Convection Velocity

Davies et al. [36] determined experimentally the convection velocity in a subsonic jet, in which a correlation is defined by the relation

$$R_{x\tau} = \frac{\overline{u_x'(0,0)u_x'(x,\tau)}}{[u_x'(0,0)]^2} = \frac{R_{x\tau}^*}{[u_x'(0,0)]^2}. \tag{5.7}$$

From the two correlation definitions, the one with the asterisk is not normalized, but the other is. The measured data can be plotted as isocorrelation curves $(R_{x\tau})$ in the (x,τ) plane (Fig. 5.2), in which the slope of the broken line represents the isocorrelation convection velocity. *Davies et al.* thus obtained the convection velocity experimentally as a function of the radial distance of the jet.

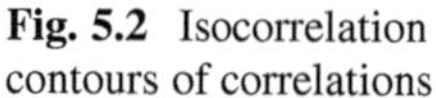

Fig. 5.2 Isocorrelation contours of correlations

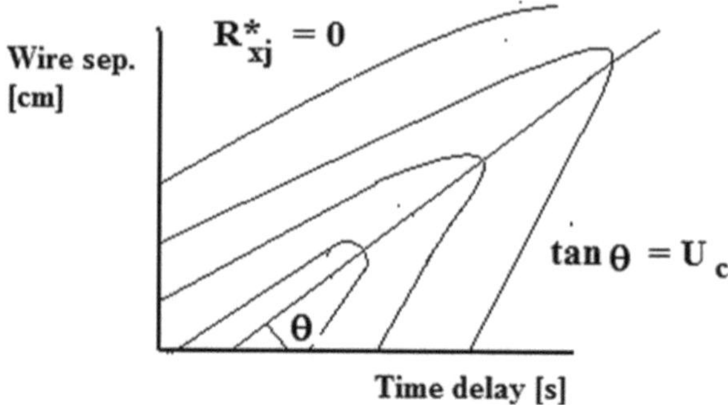

Fig. 5.3 Radial distribution of flow and convection velocity

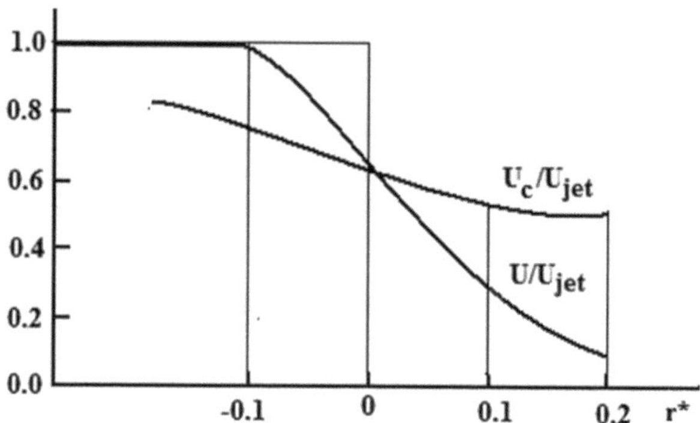

The radial distribution of the flow velocity and convection velocity, shown schematically in Fig. 5.3, shows that the convection velocity U_c changes slowly from $0.7U_{jet}$ to $0.2U_{jet}$, and it is less steep with respect to r^* than the average fluid velocity $\bar{U}$. Thus it is the usual practice for subsonic jets to assume a constant value of the convection velocity at $U_c = U_{jet}/2$.

5.3   Sound of a Convected Quadrupole

The method of *Lighthill*, as extended by Ffowcs-Williams [39], takes into account a change in the apparent volume of the sound-producing zone in the moving coordinate. Since the equation for density fluctuation with an analogy to the point source also applies to aerodynamic sound, there is an inherent difficulty in defining a volume element by an observer outside the sound-producing zone. An alternative method is adopted that appears to be simpler.

For a convected quadrupole, coordinates are given in Fig. 5.4 in which the reference point is moved from place O to distance O′, distance δ apart at time t_1, showing the relation between **y** and η at the instant t_1, when the sound was generated, to reach the observer at time t_2. Let

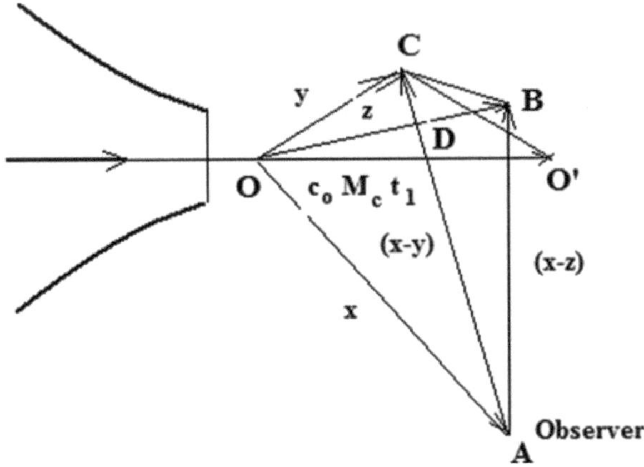

Fig. 5.4 Coordinate system for convected quadrupole in a subsonic jet

t_1 = arbitrary time when the sound wave originates at **y**,
t_1' = arbitrary time when the sound wave originates at **z**,
t_2 = time when the sound wave from **y** at time t_1 reaches the observer,
t_2' = time when the sound wave from **z** at time t_1' reaches the observer,
$\Delta t = t_2 - t_1 = |\mathbf{x} - \mathbf{y}|/c_o$,
$\Delta t_I = t_2' - t_1 = |\mathbf{x} - \mathbf{z}|/c_o$,
$\tau^* $ = time delay at observer = $t_2' - t_2$,
τ = time delay at observer = $t_1' - t_1$,
$\mathbf{U}_c$ = convection velocity,
δ = separation vector in a fixed coordinate,
$\mathbf{L}_s$ = separation vector in a moving coordinate.

Now at time t_1, the reference has moved from O to O'. Thus,

$$\mathbf{y} = \eta + c_o \mathbf{M}_c t_1 = \eta + c_o \mathbf{M}_c t_2 - \mathbf{M}_c |\mathbf{x} - \mathbf{y}|. \tag{5.8}$$

Further, one can also derive the following relations:

$$\delta = \mathbf{L}_s + c_o \mathbf{M}_c \tau = \mathbf{L}_s + c_o \mathbf{M}_c \tau^* - \mathbf{M}_c \left(|\mathbf{x} - \mathbf{y}| - |\mathbf{x} - \mathbf{z}| \right), \tag{5.9}$$

$$\tau = \tau^* + \frac{|\mathbf{x} - \mathbf{y}| - |\mathbf{x} - \mathbf{z}|}{c_o}, \tag{5.10}$$

$$|\mathbf{x} - \mathbf{y}| - |\mathbf{x} - \mathbf{z}| = CD = BC \cos(BCD) = BC \frac{\mathbf{x} - \mathbf{y}}{|\mathbf{x} - \mathbf{z}|}. \tag{5.11}$$

From (5.9)–(5.11),

$$\delta = \mathbf{L}_s + c_o \mathbf{M}_c \left[\tau^* + \frac{\delta}{c_o} \frac{(\mathbf{x} - \mathbf{y})}{|\mathbf{x} - \mathbf{y}|} \right]. \tag{5.12}$$

Now from the *Doppler principle*, $\mathbf{r} = |\mathbf{x} - \mathbf{y}|$ is replaced by the relation

$$\mathbf{r}' = \mathbf{r}(1 - M_c \cos(BCD)) = |\mathbf{x} - \mathbf{y}| \left[1 - M_c \frac{\mathbf{x} - \mathbf{y}}{|\mathbf{x} - \mathbf{y}|}\right]. \tag{5.13}$$

Since $\partial/\partial r' = -c_o^{-1}\partial/\partial t$, $|\mathbf{x} - \mathbf{y}| = \sqrt{\sum(x_i - y_i)^2}$, and

$$\frac{dr}{dx_i} = \frac{(x_i - y_i)}{|\mathbf{x} - \mathbf{y}|}\left[1 - M_c \frac{\mathbf{x} - \mathbf{y}}{|\mathbf{x} - \mathbf{y}|}\right], \tag{5.14}$$

we can write, in analogy to (3.24), the far field solution of the noise-producing term of a point quadrupole as

$$\frac{\partial^2}{\partial x_i \partial x_j}\left(\frac{T_{ij}}{r'}\right) = \frac{(x_i - y_i)(x_j - y_j)}{|(\mathbf{x} - \mathbf{y}|^2 \left[1 - M_c \frac{\mathbf{x}-\mathbf{y}}{|\mathbf{x}-\mathbf{y}|}\right]^2} \frac{\partial^2}{\partial r'^2}\left(\frac{T_{ij}}{r'}\right)$$

$$= \frac{(x_i - y_i)(x_j - y_j)}{|\mathbf{x} - \mathbf{y}|^2 \left[1 - M_c \frac{\mathbf{x}-\mathbf{y}}{|\mathbf{x}-\mathbf{y}|}\right]^2} \frac{1}{c_o^2 r'}\frac{\partial^2 T_{ij}}{\partial t^2}$$

$$= \frac{(x_i - y_i)(x_j - y_j)}{|\mathbf{x} - \mathbf{y}|^3 \left[1 - M_c \frac{\mathbf{x}-\mathbf{y}}{|\mathbf{x}-\mathbf{y}|}\right]^3} \frac{1}{c_o^2}\frac{\partial^2 T_{ij}}{\partial t^2}. \tag{5.15}$$

Further, the expressions for the density and intensity of sound are

$$\rho(\mathbf{x} - t_2) - \rho_o = \frac{1}{4\pi c_o^2}\int \frac{(x_i - y_i)(x_j - y_j)}{c_o^2[|\mathbf{x} - \mathbf{y}| - M_c(\mathbf{x} - \mathbf{y})]^3}\frac{\partial^2 T_{ij}}{\partial t^2}(\eta, t_1)d\Omega(\eta) \tag{5.16}$$

$$I(\mathbf{x}) = \frac{1}{16\pi^2 c_o^5 \rho_o}\int\int \frac{(x_i - y_i)(x_j - y_j)(x_k - z_k)(x_l - z_l)}{[|\mathbf{x} - \mathbf{y}| - M_c(\mathbf{x} - \mathbf{y})]^3 [\mathbf{x} - \mathbf{z}| - M_c(\mathbf{x} - \mathbf{z})]^3}$$

$$\times \frac{\partial^2 T_{ij}}{\partial t^2}(\eta, t_1)\frac{\partial^2 T_{kl}}{\partial t^2}(\eta + \mathbf{L}_s, t_1')d\Omega(\eta)d\Omega^*(\mathbf{L}_s). \tag{5.17}$$

First we define

$$T_{ijkl}(\eta, \mathbf{L}_s, \tau) = \overline{T_{ij}(\eta, t_1)T_{kl}(\eta + \mathbf{L}_s, t_1')}. \tag{5.18}$$

Further, since $d\Omega(\eta)$ is the *volume element* to calculate the intensity from a unit volume, whereas $d\Omega^*(\mathbf{L}_s)$ is the *correlation volume* in a moving coordinate system, and further, since the analysis is valid for an analogous point source, therefore $d\Omega(\eta)$ and $d\Omega^*(\mathbf{L}_s$ are replaced by the volumes in fixed coordinates, respectively. Thus by defining a correlation in a fixed coordinate

$$R_{ijkl}(\mathbf{y}, \delta, \tau) = T_{ijkl}(\eta, \mathbf{L}_s, \tau) = \overline{T_{ij}(\eta, t_1)T_{kl}(\eta + \mathbf{L}_s, t_1')}, \tag{5.19}$$

and from (5.12) the derivatives in two coordinates are related by the expression

$$\frac{\partial}{\partial \tau} R_{ijkl}(\mathbf{y}, \delta, \tau) = \frac{\partial}{\partial \tau} T_{ijkl}(\eta, \mathbf{L}_s, \tau) \frac{\partial \mathbf{L}_s}{\partial \delta}$$

$$= \left[1 - \mathbf{M}_c \frac{(\mathbf{x} - \mathbf{y})}{|\mathbf{x} - \mathbf{y}|} \right] \frac{\partial}{\partial \tau} T_{ijkl}(\eta, \mathbf{L}_s, \tau).$$

Further, for many cases $(\mathbf{x} - \mathbf{y}) = (\mathbf{x} - \mathbf{z})$. Therefore, the final expression for the intensity of sound is

$$I(\mathbf{x}) = \frac{1}{16\pi^2 c_o^5 \rho_o} \frac{(x_i - y_i)(x_j - y_j)(x_k - z_k)(x_l - z_l)}{|\mathbf{x} - \mathbf{y}|^6 \left[1 - \mathbf{M}_c \frac{(\mathbf{x}-\mathbf{y})}{|\mathbf{x}-\mathbf{y}|} \right]^5} \frac{\partial^4 R}{\partial \tau^4}(\mathbf{y}, \delta, \tau)$$

$$d\Omega(\mathbf{y}) d\Omega(\eta)$$

$$= \frac{c_o}{\rho_o} B_{(\tau^*=0)}, \tag{5.20}$$

where

$$B_{(\tau^*=0)} = \frac{1}{16\pi^2 c_o^6} \frac{(x_i - y_i)(x_j - y_j)(x_k - z_k)(x_l - z_l)}{|\mathbf{x} - \mathbf{y}|^6 \left[1 - \mathbf{M}_c \frac{(\mathbf{x}-\mathbf{y})}{|\mathbf{x}-\mathbf{y}|} \right]^5} \frac{\partial^4 R}{\partial \tau^4}(\mathbf{y}, \delta, \tau) d\Omega(\mathbf{y}) d\Omega(\eta).$$

By dimensional analysis, as was done earlier in Sect. 4.1, it is possible to show that the intensity of sound is given by

$$I \sim \frac{\rho_o U_{\text{jet}}^4 v^4}{c_o^5 r^2 (1 - M_c \cos \theta)^5} d\Omega d\Omega^*, \tag{5.21}$$

where θ is the angle made by the observer with the jet axis. Now let $U_c \sim U_{\text{jet}}$, frequency $v \sim U_c/D_o$, $d\Omega \sim D_o^3$, $d\Omega^* \sim D_o^3$, and acoustic power $P \sim Ir^2$. Thus the intensity of sound is

$$I \sim \frac{\rho_o U_c^8 D_o^2}{r^2 c_o^5 (1 - M_c \cos \theta)^5} \tag{5.22}$$

and the *acoustic power* is

$$P \sim \frac{\rho_o U_c^8 D_o^2}{c_o^5 (1 - M_c)^5}. \tag{5.23}$$

At low subsonic speeds, $M_c \ll 1$, the results of (5.16) and (5.17) converge to those given in Table 4.1. For high supersonic convections, however, when $M_c \cos \theta \gg 1$,

$$I \sim \frac{\rho_o U_c^3 D_o^2}{r^2 \cos^5 \theta} \tag{5.24}$$

and

$$P \sim \rho_o U_c^3 D_o^2. \tag{5.25}$$

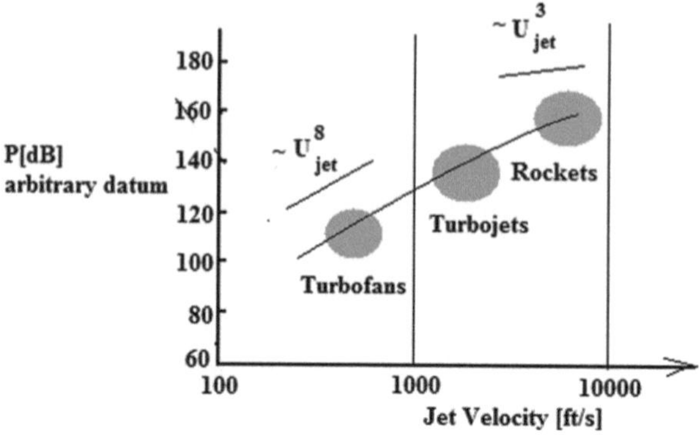

Fig. 5.5 Variation of acoustic power levels

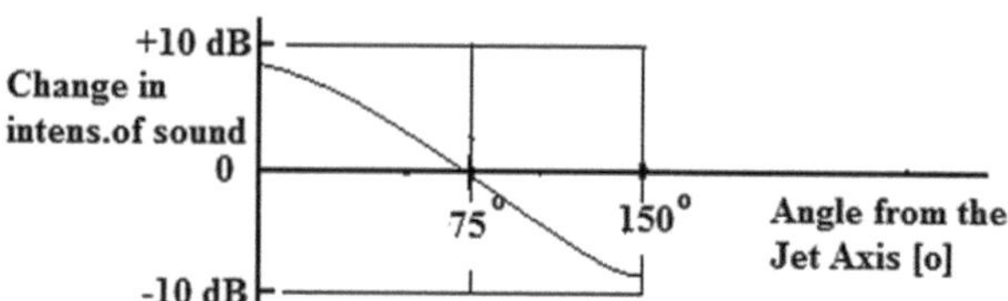

Fig. 5.6 Directional distribution of sound radiation from airjets at $300\,\mathrm{ms}^{-1}$, in decibels relative to $\Omega = 75°$

The overall power level was given by Chobotov and Powell, and reproduced [39] after suitably adjusting to account for the diameter. This is shown in Fig. 5.5.

To obtain the acoustic power in a moving coordinate, the value of the integral, similar to (4.17) for a convected quadrupole, is

$$P = \int_{\phi=0}^{2\pi} \int_{\theta=0}^{\pi} \frac{x_i x_j x_k x_l}{x^4 (1 - M_c \cos\theta)^5} \sin\theta \, d\phi \, d\theta, \tag{5.26}$$

which needs to be evaluated.

Ffowcs-Williams' [39] theory of moving quadrupoles describes satisfactorily one of the most noticeable features of jet noise, namely, its mark directionality, which is greater in intensity at small angles θ between the jet direction and the direction of the emission, a variation that becomes more pronounced as jet velocity increases. For example, Fig. 5.6, which is based schematically on Fig. 13 of Lighthill [61], shows the directional distribution of sound for stationary cold jets at $300\,\mathrm{ms}^{-1}$, with the curve representing the factor $(1 - M_c \cos\theta)^{-1}$ using an average convection velocity of $U_c = 0.5 U_{\mathrm{jet}}$. The measured directional distributions are in general in good agreement, except that they fall off somewhat and become slower behind the nozzle exit, probably because other sources of sound are additionally detectable behind the nozzle exit. These include the additional fields of the weaker peripheral eddies with lower velocities of convection and also some reverberation of much higher intensity around transmitted forwards.

Indeed, the theory indicates only how the directional distribution should change with jet velocity; it does not say anything about the preferred orientation of the

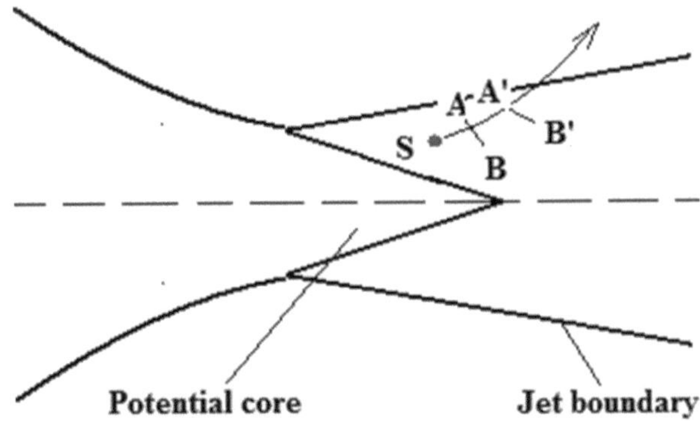

Fig. 5.7 Schematic sketch of refraction of noise

quadrupoles. *Lighthill* argued that, in any region of large mean shear, there would be some predominance of lateral quadrupoles with directional peaks at 45° to the direction of the flow.

The previous results of amplification, as obtained by *Ffowcs-Williams*, postulate very large decay times for eddies. The convection amplification factor $(1 - M_c \cos\theta)^{-5}$ contained in the sound power from the slice of jet with frequency v appears here as a change in the radial coordinate as $(1 - M_c \cos\theta)^{-6}$ and a *Doppler shift* of the spacing $(1 - M_c \cos\theta)$. A progressive redirection of frequency in the jet flow direction pushes the peak slightly, giving a somewhat smaller effective shift,

$$\left[(1 - M_c \cos\theta)^2 + \alpha^2 M_c^2 \right]^{-5/2}, \tag{5.27}$$

instead of $(1 - M_c \cos\theta)^{-5}$, where α takes care of the *decay time of eddies*. For unheated air jets $\alpha = 0.33$, and for turbojets $\alpha = 0.55$.

Figure 4.4 shows the basic subsonic jet noise pattern for unconvected quadrupoles, but including the shear and the self noise. However, because of convection, (4.28) can, therefore, be written in the following modified form:

$$I \sim \frac{A + B(\cos^4\theta + \cos^2\theta)/2}{\left[(1 - M_c \cos\theta)^2 + \alpha^2 M_c^2 \right]^{5/2}}. \tag{5.28}$$

It is further well known from theory and experiments that sound waves are refracted by the velocity gradient, that is, by wind from a subsonic jet. To see why the ray is bent (Tam and Auriault) [110], we need to consider only the propagation of the wave front AB (Fig. 5.7). Point A moves at a speed equal to the local sound speed plus the local flow velocity of the jet. This is also true of point B if the flow velocity at B is higher. Therefore, as the wave front propagates, it becomes tilted, even for an isothermal jet. Obviously this effect of refraction is even more pronounced in hot jets, where the sound speed at B is higher than at A. One of the important consequences of mean flow refraction is that less sound can be radiated in the direction of the jet flow. This creates a relatively quiet region around the jet axis

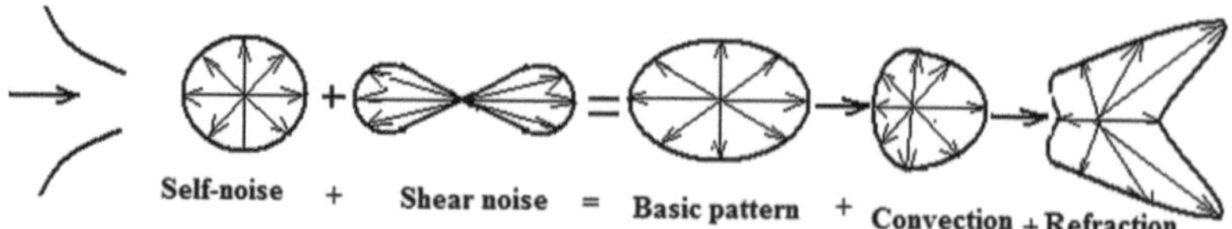

Fig. 5.8 Jet noise pattern under influence of convection and refraction

commonly known as the *cone of silence*. On the other hand, if the jet is very cold, the same argument would lead to the conclusion that the sound rays may bend toward the jet axis, leading to a large increase in the jet intensity there.

In the case of plane waves in a semi-infinite pane flow the flow angle of refraction and the refracted intensity can be predicted fairly accurately. There exists a minimum refraction angle given by

$$\sec \theta = \frac{U_{\text{jet}}}{c_{\text{jet}}} + \frac{c_i}{c_{\text{jet}}}, \tag{5.29}$$

which corresponds to a grazing incidence of $(\theta_{\text{jet}} \to 0)$. Note that subscript i stands for incidence.

Thus, combining the effect of convection and refraction, the total radiation pattern of a subsonic jet noise is given in Fig. 5.8.

5.4 Exercises

5.4.1 An electric railway engine with a shrill whistle at a characteristic frequency of 120 Hz roars past a railway station at 80 km/h^{-1}, where the speed of sound is 330.0 ms^{-1} when the outside temperature is a comfortable 300 K. What are the wavelength and frequency of the whistle sound of the approaching and leaving engine for a person standing on the wayside railway platform? What would be the values for these on a cold day when the outside temperature is freezing?

5.4.2 What are *isocorrelation curves* and *convection velocity*?

5.4.3 What are the radial flow and convection velocity distributions in the various regions of a subsonic jet?

5.4.4 How does the acoustic power change with jet velocity?

5.4.5 Explain the refraction of jet noise and the effect of convection and refraction on the directional pattern of jet noise.

Chapter 6
Computational Aeroacoustics

Computational aeroacoustics (CAA) deals with the prediction of an aerodynamic noise source and its propagation numerically with the help of time-dependent equations. Some of the great books in this field are the 1976 book by Goldstein [3] and a recent book by Wagner et al. [12].

After identification of the sound sources, the noise generated in the field must be transported outside to an observer. *Computational fluid dynamics* (CFD) was designed generally to solve a near-field problem because the perturbations from the mean flow vanish quickly. Furthermore, the flow in this region is usually highly nonlinear but basically stationary, or at least the changes are slow. On the other hand, acoustics is clearly a far-field problem, in which the sound is generated locally in the aerodynamic area and passively radiated outside to an observer with a smaller exponent of decrease with radial distance. Otherwise the area where the sound is generated is aerodynamically active, the perturbations are small, and linear descriptions are usually sufficient. However, noise due to turbulence is inherently unsteady quite comparable to turbulent eddies even if spatial wavelengths are large compared with aerodynamic ones by an order of the reciprocal of the Mach number.

Aerodynamic noise occurs basically because of three different phenomena. The first one is due to a fluctuating mass flow rate, as the noise that can happen from sirens or in exploding burning droplets in a combustion chamber (*combustion noise*) as a result of the chemical reactions and the subsequent introduction of energy into the flow (*entropy noise*).

The second one is deterministic *impulsive noise*, which is a result of moving surfaces in nonuniform flow conditions. The *displacement effect* of an immersed body in motion and the nonstationary aerodynamic loads on the body's surface, generate loads on the bodys' surface creating pressure fluctuations that are radiated as sound. This type of noise is relatively easy to extract from aerodynamic simulations because the required resolution in space and time to predict the acoustics is similar to the demands from aerodynamic computations. Aerodynamic noise arises primarily from *rotating systems* (e.g., helicopter rotors, wind turbines, cooling fans, and ventilators). If the surface moves at speeds comparable to the speed of

T. Bose, *Aerodynamic Noise: An Introduction for Physicists and Engineers*,   83
Springer Aerospace Technology 7, DOI 10.1007/978-1-4614-5019-1_6,
© Springer Science+Business Media New York 2013

sound or there is an interaction between a rotor and a stator wake, then the tonal engine components can be dominant [12]. Because the frequency of interaction between a rotor and a stator can be determined by the product of the number of blades and the revolutions per minute, analysis of the noise level is an easy way of determining problems in a particular blade row.

The third noise mechanism is the result of *turbulence* and therefore arises in nearly every engineering application. Turbulence is, by its very nature, a *stochastic phenomenon* and therefore has a broad frequency spectrum (*broadband frequency spectrum*). Interestingly enough, turbulence energy is converted into acoustic energy most efficiently in the vicinity of sharp edges (for example, at the *trailing edge of an aircraft wing*). In this case, the uncorrelated turbulent eddies flowing over the upper and lower sides of the edge must relax with each other, generating locally strong, equalizing flows that result in highly nonstationary pressure spikes. Another major source of turbulence-related sound is jet flows, in which the shear layer in the mixing zone again radiates sound in the far field [12].

The full 3-dimensional Navier–Stokes equations describe the basic phenomena, just as they do for the solution of computational aerodynamics (CAD) or CFD. However, using the CAD techniques for CAA applications may have a number of difficulties, and it is worth examining the use of these methods for applications in aeroacoustics.

According to *Tam et al.* [111], "most current computational fluid dynamics (CFD) finite difference schemes are designed for the solution of time-independent problems." In the formulation of these schemes, requirements of consistency and numerical stability are imposed. By invoking the Lax equivalence theorem, the convergence of these schemes is then assured. The quality of CFD schemes is generally ranked by the order of (Taylor series) truncation. It is expected that a fourth-order scheme is better than a second-order scheme and so on. For time-independent problems, the foregoing criteria are quite sufficient. But for time-dependent problems, a consistent, stable, and convergent high-order scheme does not guarantee a good-quality numerical wave solution. A simple analysis of the linearized compressible Euler equations reveals that in a uniform mean flow, partial differential equations support three types of waves: acoustic, entropy, and vorticity waves. Acoustic waves are isotropic, nondispersive, and nondissipative and propagate at the speed of sound. Entropy and vorticity waves are nondispersive, nondissipative, and highly directional. They propagate in the direction of the mean flow at the same speed as the flow. On the other hand, many current CFD codes are dispersive, anisotropic, and highly dissipative (sometimes artificial terms are added to improve stability). It is, therefore, necessary to discuss numerical nondissipative schemes.

The various acoustic wave propagation properties and turbulent flow properties [linear versus nonlinear, far field versus near field, time dependent versus (quasi-) stationary, large versus small spatial scale] obviously necessitate different tradeoffs regarding computational schemes. No single known scheme scores best on all the possible requirements of CFD and CAA. For applications with complex, inhomogeneous flows and flow-induced noise radiation, the most promising and commonly used numerical technique is to adopt a *hybrid method* such as the coupling of the *Reynolds-averaged Navier–Stokes approach* (RANS) and *large eddy simulation* (LES) for *detached eddy simulation (DES)*, in which the sound generation due to aerodynamic causes is more or less decoupled from the acoustic transport process to the far field, allowing for the use of tailored algorithms for both tasks. This decoupling obviously results in an arbitrary combination of a *sound generation method* with another *sound transport method*.

Some of the methods are as follows:

(a) Direct numerical simulation (DNS), which can be considered the most exact technology. The complete, fully coupled compressible Navier–Stokes and Euler equations are solved in the domain of interest for the unsteady combined flow and acoustic field from the effective area down to the far-field observer. However, it requires tremendous computational resources because the flow and acoustics represent a multiscale problem. The small acoustic perturbations are drowned out by numerical errors of the much larger aerodynamic forces. Space and time derivative requirements for the aerodynamic data combined with the large distances up to an observer in the far field give rise to a ridiculously high number of cells and time steps.

(b) Computational transport by solving simple equations, such as *linearized Euler equations (LEE)* (LEEs) or simply the wave equation. "Because these are both linear, many problems encountered with a full set of equations do not occur, and consequently the discretization schemes can be highly tuned to reach the desired low level of dispersion and diffusion errors. On the boundary between the CAA domain and the CFD computation, the CFD solution is used as a boundary condition for the CAA simulation" [12].

(c) Analytical transport techniques employ an integrated form of the relevant acoustic propagation equation, like *Kirchoff's surface integral equation*, which is explained subsequently in a separate section. "In this case the sound pressure at an observer at a specific point of time is computed by an integration of source terms along a surface – either a physical one or surrounding the aerodynamic area. Owing to the finite speed of sound and deterministic relation between emission and observer time of a signal, there has to be some kind of interpolation of the data at least on one side. In case some parts of the integration surface or volume are moving at transonic speeds, the integral becomes highly singular because of the *Doppler effect*, which leads to difficulties regarding the stability of the procedure" [12].

6.1 Numerical Nondissipative Schemes

According to *Roe* [97], the following distinctions are often made between aerodynamics and aeroacoustics:

(a) Most aerodynamic codes to date have been concentrated on predicting steady flows. Although these are often designed as the limit of a time-dependent flow, the *time-marching techniques* are designed to eliminate all transients as quickly as possible; an aeroacoustic code would need to preserve these.

(b) Because of the acoustic nature of acoustic waves, whose amplitudes may be of the same order as the truncation errors present in the underlying nonlinear flows, it was feared that aeroacoustic solutions would be corrupted by noise.

(c) Boundary conditions may be harder to formulate, and failure to impose proper boundary conditions in an acoustic problem may lead to obtaining the wrong solution.

(d) Aerodynamically generated sound is governed by a nonlinear process, and the technologically important applications are for large Reynolds-number-averaged turbulent flows in steady state, whereas for CAA purposes, time-dependent flow fluctuations are needed [66].

As a typical scenario, it was suggested by Roe [97] that computations may be done in two stages. In the first stage, an aerodynamic code with fixed boundary conditions is run to establish a steady flow behavior, followed by an acoustic code with small, periodic boundary conditions, until a steady acoustic pattern evolves. In this connection it is argued (Thomas and Roe [114]) that schemes are needed that are inherently free of dissipation, for which the "leapfrog scheme" and an "upwind form" of the scheme are candidates and the standard CFD schemes are found to be inadequate even for the simplest aeroacoustics problems. In addition, a very relevant parameter is the number of mesh points, N, required per wavelength to hold dissipation and dispersion within acceptable bounds. It is said that to achieve 1% accuracy in the position of the wave will require N to be on the order of 20–25. In general, eliminating dissipation is easy; the classic leapfrog method does the job. In addition to keeping N small, a high-order upwind scheme is proposed. Superimposed on the steady flow we have a pressure source oscillating at a specified frequency with an amplitude of 10^{-8} that of the inlet pressure, which is set at the aft end of the nozzle. The leapfrog discretization for scalar advection ($u_t + au_x = 0$, $a > 0$) can be expressed as

$$u_j^{n+1} = u_j^{n-1} - r\left(u_{j+1}^n - u_{j-1}^n\right), r = c\frac{\Delta t}{\Delta x}, \tag{6.1}$$

where the superscript is for the time index and the subscript is for the index. However, in the scheme, the difference between numerical and exact wave speeds increases, especially if the number of computational grid points is not selected properly. For this the relevant parameter is the number N of mesh points required to hold dissipation within acceptable bounds. For most methods, to keep the phase error within 1% accuracy, N must be 20–25 per time step, and for the propagation of waves over R wavelengths, the attenuation in amplitude per wavelength on the order of 1% is $(1 - R/(vN))$ and the total number of time steps required is RN/v [114]. Thus, to remain within 1% of the wave speed at a CFL number of 0.5, at least 16 cells per wavelength are needed. Another scheme, called the *upwind leapfrog method*, for $a > 0$, with neutral stability for all CFL numbers, is

$$\frac{\left(u_j^{n+1} - u_j^n\right) + \left(u_j^{n-1} - u_{j-1}^{n-1}\right)}{2\Delta t} + a\frac{u_j^n - u_{j-1}^n}{\Delta x}, \tag{6.2}$$

which follows

$$u_j^{n+1} = (1 - 2r)\left(u_j^n - u_{j-1}^n\right) + u_{j-1}^{n-1}, \tag{6.3}$$

where $r = a\Delta t/\Delta x$ and $a > 0$. For the advection speed $a < 0$, the corresponding scheme is

$$u_j^{n+1} = (1 + 2r)\left(u_{j+1}^n - u_j^n\right) + u_{j+1}^{n-1}. \tag{6.4}$$

In this connection it is possible to estimate the minimum number of points required for spatial discretization. Although a finite difference equation can be transformed with the help of a *Fourier transform* into one in the frequency domain, it is obvious that a sinusoidal wave form requires at least five grid points to perform a simulation [111].

For the purpose of discretization of flow and acoustic equations, MacCormack schemes have been used extensively for constant spatial step size. The most popular and well-tested of the MacCormack type schemes is one with second-order accuracy with respect to time and second order in space. However, an improved scheme, with second-order accuracy in time and fourth-order accuracy is discussed by Hixon [50]. For this the forward and backward discretizations of the fluxes in a MacCormack scheme can be written as

$$\left(\frac{\partial \mathbf{F}}{\partial x}\right)^f = a_{-1}\mathbf{F}_{i-1} + a_0\mathbf{F}_i + a_1\mathbf{F}_{i+1} + a_2\mathbf{F}_{i+2} + a_3\mathbf{F}_{i+3}, \tag{6.5}$$

$$\left(\frac{\partial \mathbf{F}}{\partial x}\right)^b = a_{-1}\mathbf{F}_{i+1} + a_0\mathbf{F}_i + a_1\mathbf{F}_{i-1} + a_2\mathbf{F}_{i-2} + a_3\mathbf{F}_{i-3}. \tag{6.6}$$

Some of the two-step schemes with forward and backward steps, like *MacCormack schemes*, can work only with equal space intervals in order to have proper cancellation of numerical errors in the alternate directions.

The accuracy of MacCormack schemes is increased by increasing the number of points of the stencil and optimizing the dispersion error by adding another point $(a - 1)$ in the opposite direction to each one-sided difference. Originally, explicit MacCormack schemes were two-four, which is second-order accurate in space (with constant space interval) and second-order in time. The two-four scheme uses the operator splitting technique with alternating the operators symmetrically in order to maintain the accuracy of second order in time and fourth order in space, and it has been extended to sixth order spatial accuracy by Bayliss et al. [23]. Thus the accuracy of MacCormack schemes is increased by increasing the number of points in the stencil and the dispersion error is optimized for these schemes by adding another point in the opposite direction to each one-sided difference.

The usual procedure to obtain first-order finite difference formulas is to write from the expansion of *Taylor series* and truncate at the term depending on the specified number of points desired to derive finite difference formulas. The coefficients of these are given in the first four rows in the following table. Tam and Webb [111] have constructed difference schemes in a slightly different way. For this the finite difference formulas for forward and backward schemes have been written as a general formula:

$$\frac{\partial \mathbf{F}}{\partial x} = \frac{1}{\Delta x} \sum_{j=-N}^{M} a_j \mathbf{F}(x + j\Delta x). \qquad (6.7)$$

First we get the Fourier transform and the inverse of the function

$$\mathbf{F}^*(\omega) = \frac{1}{2\pi} \int_{-\infty}^{\infty} \mathbf{F}(x) \exp^{-i\omega x} dx \qquad (6.8)$$

$$\mathbf{F}(\omega) = \int_{-\infty}^{\infty} \mathbf{F}^*(x) \exp^{i\omega x} dx, \qquad (6.9)$$

and thus the Fourier transform of the derivative equation becomes

$$i\omega \mathbf{F}^* = \left(\frac{1}{\Delta x} \sum_{j=-N}^{M} a_j \exp^{i\omega j \Delta x} \right) \mathbf{F}^*. \qquad (6.10)$$

Comparing both sides, the effective wave number of the Fourier transform of the finite difference scheme is

$$\omega^* = \frac{-i\omega}{\Delta x} \sum_{j=-N}^{M} a_j \exp^{i\omega j \Delta x}, \qquad (6.11)$$

where $\omega^* \Delta x$ is a periodic function of $\omega \Delta x$ with period 2π. Now to make sure that the Fourier transform of the finite difference scheme is a good approximation within a given number of points in the range of the wavelength of interest, it is required that a_j be chosen so as to minimize the integrated error. For instance, for $N = M = 3$, obtaining an accuracy within $(\Delta x)^4$, the coefficients have the values

$$a_0 = 0, a_{-1} = a_1 = 0.79926643, a_{-2} = a_2 = -0.18941314,$$

$$a_{-3} = a_3 = 0.02651995. \qquad (6.12)$$

The preceding coefficients for a *dispersion-relation-preserving* (DRP) scheme are given by Tam and Webb [111]. Comparison of a DRP scheme with standard schemes has been done by Owis and Balakumar [84]. It appears that there is good agreement between the two-four scheme and a DRP scheme for a grid size of 17 points per wavelength in the axial direction. As the amplitude of the externally applied inlet disturbance grows, the linear stability results deviate from the results obtained using a MacCormack scheme because the linear stability theory is not valid for large disturbances. As the grid size in the axial direction is reduced to nine points per wavelength, the two-four scheme has a higher dispersion error compared to optimized DRP schemes, and the errors become worse as the number of points per wavelength is further reduced. However, the dispersion error for two-six schemes with the same grid size is lower than that of two-four schemes, but the amplitude of the axial velocity disturbance is higher than that of the DRP scheme for a fine grid.

Finally, tests with a four-six scheme (an optimized version of the two-six scheme) showed better results than both the two-four scheme and two-six schemes, but DRP schemes had the best results.

Bailly and Juve [18] discretized the space derivative in the LEE by using seven-point discretization and constant step size in either direction ($\Delta x = \Delta y$) to get (Sect. 6.2)

$$\frac{\partial \mathbf{U}}{\partial t} - \sum_{l=3}^{3} a_l \left(\mathbf{E}_{i+1,j} + \mathbf{F}_{i,j+1} \right) - \mathbf{H}_{i,j} + \mathbf{S}_{i,j} \tag{6.13}$$

by requiring that the effective wave number provided by the finite difference scheme be a close approximation to the actual wave number. For this a optimized fourth-order scheme is used. The authors point out that seven mesh points per wavelength are necessary for the DRP scheme of Tam and Webb [111]; the standard six-order scheme requires ten mesh points per wavelength. In some cases it is necessary to remove spurious numerical oscillations due to nonlinearities or mismatches at the boundary, and these were removed by an artificial selective damping discussed by the authors. The time integration is performed by a four-step Runge–Kutta algorithm.

6.2 Turbulence Flow Equations

In the previous chapters, it was shown that noise is linked to the fluctuations of mass, force, and turbulence. These are described by the full *Navier–Stokes equations* by including viscosity, heat conductivity, and energy transfer terms like *radiation*. For our case, we will exclude the energy equation and mass production and volume force term from our consideration of the *basic governing equations*.

6.2.1 Basic Governing Equations

The basic governing equations excluding the energy equation and volume source terms like mass production or volume force are written as follows:

$$\text{Continuity}: \frac{\partial \rho}{\partial t} + \frac{\partial (\rho u_j)}{\partial x_j} = 0, \tag{6.14}$$

$$\text{Momentum}: \frac{\partial \rho u_i}{\partial t} + \frac{\partial (\rho u_i u_j)}{\partial x_j} = -\frac{dp}{dx_i} + \nabla \bullet \tau, \tag{6.15}$$

where

$$\tau_{ij} = \mu \left[\left(\frac{\partial u_i}{\partial x_j} + \frac{\partial u_j}{\partial x_i} \right) - \frac{2}{3} \frac{\partial u_j}{\partial x_j} \delta_{ij} \right] \tag{6.16}$$

in which μ is the (laminar) *viscosity coefficient* and δ is the *Kronecker delta* with values $\delta_{ii} = 1$ and $\delta_{ij} = 0$.

6.2.2 Turbulence Flow Equations

The flow equations in the previous sections, however, do not include turbulence terms, which are so important for the generation of *aerodynamic noise*. In fact, the solution of the *turbulence flow equations* remains the most difficult unsolved problem to date, and the method for solving it is to take the method of *statistical turbulence* and then use *semiempirical formulas*.

The first try in that direction was by the French scientist *Boussinesque* (1877). According to him, flow properties are split into *time-averaged* (independent of time) and *time-dependent* components as follows:

$$\rho = \bar{\rho} + \rho'(t), \tag{6.17}$$

$$p = \bar{p} + p'(t), \tag{6.18}$$

$$u_i = \bar{u}_i + u_i'(t). \tag{6.19}$$

These properties can also be written in terms of the *mass-weighted average*. Accordingly,

$$\overline{\rho u_i'} = \overline{(\bar{\rho} + \rho'')u_i'} = \overline{\bar{\rho} u_i'} + \overline{\rho'' u_i'} = \bar{\rho}\overline{u_i'} + \overline{\rho'' u_i'} = 0 \tag{6.20}$$

by definition, and thus,

$$\overline{\rho u_i''} = 0, \tag{6.21}$$

whereas

$$\overline{u_i''} \neq 0. \tag{6.22}$$

Note that

$$\overline{u_i'} = 0 \tag{6.23}$$

but

$$\overline{\rho u_i'} \neq 0. \tag{6.24}$$

Now,

$$\overline{\rho u_i'} = \overline{(\bar{\rho} + \rho'')u_i'} = \overline{\bar{\rho} u_i'} + \overline{\rho'' u_i'} = \bar{\rho}\overline{u_i'} + \overline{\rho'' u_i'} = 0. \tag{6.25}$$

For average calculations, very often it is assumed that

$$[\bar{a}] \cdot [\bar{b}] = [\overline{ab}]. \tag{6.26}$$

It is also estimated that $u_i' \approx u_i''$ (*boudary-layer approximation!*).

Now the basic governing *continuity* and *momentum equations* of the previous section can be written for turbulent flow in a Cartesian coordinate system as follows:
Continuity:

$$\frac{\partial \rho}{\partial t} + \frac{\partial}{\partial x_k}(\rho u_k) = 0, \ k = 1 - 3. \tag{6.27}$$

Splitting each variable in terms of time-averaged and time-dependent components, the preceding equation can be written as

$$\frac{\partial}{\partial t}(\bar{\rho}) + \frac{\partial}{\partial x_k}(\rho\tilde{u}_k + \rho u'_k) = 0. \tag{6.28}$$

By time averaging, since

$$\bar{\rho}'' = 0 \,,\, \overline{\rho\tilde{u}_k} = \bar{\rho}\tilde{u}_k, \tag{6.29}$$

we get

$$\frac{\partial}{\partial x_k}(\bar{\rho}\tilde{u}_k) = 0. \tag{6.30}$$

Let us define a rate of change (*substantive differential*) in a moving coordinate with the flow

$$\frac{D}{Dt} = \frac{\partial}{\partial t} + \tilde{u}_k\frac{\partial}{\partial x_k}. \tag{6.31}$$

The *continuity equation* can also be written as

$$\frac{D\bar{\rho}}{Dt} + \rho\frac{\partial\tilde{u}_k}{\partial x_k} = 0. \tag{6.32}$$

Note that in (6.30) and (6.32),

$$\frac{\partial\rho}{\partial t} = 0. \tag{6.33}$$

Now subtracting (6.32) from (6.30), we get

$$\frac{\partial\rho''}{\partial t} + \frac{\partial}{\partial x_k}(\rho''\tilde{u}_k + \rho u'_k) = 0. \tag{6.34}$$

Momentum:

$$\frac{\partial}{\partial t}(\rho u_i) + \frac{\partial}{\partial x_k}(\rho u_i u_k) = -\frac{\partial p}{\partial x_k} + \frac{\partial\tau_{ik}}{\partial x_k}, \tag{6.35}$$

where the *shear stress* is given by

$$\tau_{ik} = -\frac{2}{3}\bar{\mu}\frac{\partial u_k}{\partial x_k} + \bar{\mu}\left(\frac{\partial u_i}{\partial x_k} + \frac{\partial u_k}{\partial x_i}\right). \tag{6.36}$$

Once again (6.35) is expanded in terms of time-averaged and time-dependent parts as follows:

$$\frac{\partial}{\partial t}\left[\rho(\tilde{u}_i + u'_i)\right] + \frac{\partial}{\partial x_k}\left[\rho(\tilde{u}_i\tilde{u}_k + u'_i\tilde{u}_k + u'_k\tilde{u}' + u'_i u'_k)\right] = -\frac{\partial\bar{p}}{\partial x_i} - \frac{\partial p''}{\partial x_i} + \frac{\partial\bar{\tau}_{ik}}{\partial x_k} + \frac{\partial\tau''_{ik}}{\partial x_k}. $$

$$\tag{6.37}$$

By time averaging and noting that

$$\overline{\rho u'_j} = 0 \; ; \; \overline{p''} = 0 \; ; \; \overline{\tau''_{ij}} = 0 \tag{6.38}$$

we get

$$\frac{\partial}{\partial t}\left(\bar{\rho}\tilde{u}_i\right) + \frac{\partial}{\partial x_k}\left(\rho\tilde{u}_i\tilde{u}_k\right) = -\frac{\partial\bar{p}}{\partial x_i} + \frac{\partial}{\partial x_k}\left(\bar{\tau}_{ik} - \overline{\rho u'_i u'_k}\right). \tag{6.39}$$

Herein $-\overline{\rho u'_i u'_k}$ is called the *Reynolds stress* due to the turbulent motion of the fluid given by the relation

$$\tau_t = -\bar{\rho}\begin{bmatrix} \overline{u'^2} & \overline{u'v'} & \overline{u'w'} \\ \overline{u'v'} & \overline{v'^2} & \overline{v'w'} \\ \overline{u'w'} & \overline{v'w'} & \overline{w'^2} \end{bmatrix}. \tag{6.40}$$

The total shear stress is the sum of *laminar shear stress*, (6.16), and *turbulent shear stress*. While the former can be computed easily from the local velocity gradient, the latter depends both on the local velocity gradient and the time-averaged value of the product of velocity perturbations. While the proportionality constant for the shear stress in laminar flow, the *dynamic viscosity coefficient*, is a material property, there is no such material property for a turbulent flow is available to link the local shear stress with local values of the velocity perturbation. Further, while the solution of the shear stress term strictly for a *boundary layer type of flow* has been attempted quite successfully with the help of semiempirical relations, this was not flow in the case of general turbulent flows. In view of this, methods were developed from the basic continuity and momentum equations that are valid with a very limited number of universal constants.

6.2.3 Two-Equation Model

As explained by Jones and Launder [52], "we take the turbulent viscosity to be determined uniquely by the local values of the density ρ, *turbulence kinetic energy*, k, and *turbulence length scale*, l. Thus for a dimensional homogeneity" we write

$$\mu_t = c_\mu \rho k l. \tag{6.41}$$

One- or two-equation models are equations that supplement the continuity and momentum equations. In the momentum equation for turbulent flow, the *turbulent viscosity coefficient*, μ_t, and the so-called *eddy diffusity coefficient*, ε_m, are related by the equation

$$\mu_t = \bar{\rho}\varepsilon_m, \tag{6.42}$$

where c_μ is a constant.

From the momentum equation one can define the *turbulent kinetic energy equation* from the relation

$$k = \frac{1}{2}\sum_i (\overline{u_i'^2}) \tag{6.43}$$

and the *turbulent energy dissipation equation*

$$\bar{\rho}\varepsilon \equiv \sum\sum \overline{\left(\frac{\partial u_i'}{\partial x_i}\right)^2}. \tag{6.44}$$

There are several two-equation models. Three of the more popular and widely used models are the k and ε model of Jones and Launder [52], the k and ω model of Wilcox, $\omega = (vorticity)$, and the SST model of Menter, which blends both models. Of the three, the (k, ε) model is the most widely used and will be discussed here.

The *eddy diffusivity* is now

$$\varepsilon_m = \frac{\mu_t}{\rho} = c_\mu \frac{k^2}{\varepsilon}. \tag{6.45}$$

Now the k and ε equations are as follows:

k equation:

$$\frac{Dk}{Dt} = \frac{\partial}{\partial x_k}\left[\left(\nu_t + \frac{\varepsilon_m}{\sigma_k}\right)\frac{\partial k}{\partial x_k}\right] + \varepsilon_m\left(\frac{\partial \bar{u}_i}{\partial x_j} + \frac{\partial \bar{u}_j}{\partial x_i}\right)\frac{\partial \bar{u}_i}{\partial x_j} - \varepsilon; \tag{6.46}$$

ε equation:

$$\frac{D\varepsilon}{Dt} = \frac{\partial}{\partial x_k}\left[\left(\nu + \frac{\varepsilon_m}{\sigma_\varepsilon}\right)\frac{\partial \varepsilon}{\partial x_k}\right] + c_{\varepsilon 1}\frac{\varepsilon}{k}\varepsilon_m\left(\frac{\partial \bar{u}_i}{\partial x_j} + \frac{\partial \bar{u}_j}{x_i}\right)\frac{\partial \bar{u}_i}{\partial x_j} - c_{\varepsilon 2}\frac{\varepsilon^2}{k}. \tag{6.47}$$

The value of the constants here are

$$c_\mu = 0.09, c_{\varepsilon 1} = 1.44, c_{\varepsilon 2} = 1.92, \sigma_k = 1.0, \quad \text{and} \quad \sigma_\varepsilon = 1.3. \tag{6.48}$$

In order to provide predictions of the flow for low *Reynolds numbers* within the *viscous layer* adjacent to the wall, the c "will become dependent on the Reynolds number of turbulence", and "further terms have to be added to account for the fact that the dissipation processes are not isotropic [52]." The complete form of the k and ε equations are now as follows:

k equation:

$$\frac{Dk}{Dt} = \frac{\partial}{\partial x_k}\left[\left(\nu_t + \frac{\varepsilon_m}{\sigma_k}\right)\frac{\partial k}{\partial x_k}\right] + \varepsilon_m\left(\frac{\partial \bar{u}_i}{\partial x_j} + \frac{\partial \bar{u}_j}{\partial x_i}\right)\frac{\partial \bar{u}_i}{\partial x_j} - \varepsilon - 2\mu_t\left(\frac{\partial \sqrt{k}}{\partial y}\right)^2;$$

$$\tag{6.49}$$

ε equation:

$$\frac{D\varepsilon}{Dt} = \frac{\partial}{\partial x_k}\left[\left(v + \frac{\varepsilon_m}{\sigma_\varepsilon}\right)\frac{\partial \varepsilon}{\partial x_k}\right] + c_{\varepsilon 1}f_1\frac{\varepsilon}{k}\varepsilon_m\left(\frac{\partial \bar{u}_i}{\partial x_j} + \frac{\partial \bar{u}_j}{x_i}\right)\left(\frac{\partial \bar{u}_i}{\partial x_j}\right)^2$$

$$-c_{\varepsilon 2}f_2\frac{\varepsilon^2}{k} + 2\frac{\mu}{\mu_t}\rho\left(\frac{\partial^2 u_i}{\partial x_j^2}\right);\tag{6.50}$$

and the turbulence viscosity formula becomes

$$\mu_t = c_\mu f_\mu \rho\frac{k^2}{\varepsilon},\tag{6.51}$$

where

$$f_1 = 1.0,$$

$$f_2 = 1.0 - 0.3\exp^{-R^2},$$

$$f_\mu = \exp^{-25/(1+R/50)},\tag{6.52}$$

where $R = \rho k^2/(\mu\varepsilon_m)$ may be interpreted as the *turbulence Reynolds number*. At the time of formulation of the constant f_1 were tried as a function of *Reynolds number*. But at a later time it was put equal to unity.

The (k,ε) equations are solved along with the usual fluid-dynamic equations. At the wall, normally *no-slip conditions* are used for the fluid-dynamic equations; for k and ε the following boundary conditions are imposed [52]:

$$y \to 0 : k = 0, \varepsilon = 0,\tag{6.53}$$

$$y \to \infty : u_\infty\frac{dk_\infty}{dx} = -\varepsilon_\infty u_\infty\frac{d\varepsilon_\infty}{dx} = -c_2 f_2\varepsilon_\infty^2/k_\infty.\tag{6.54}$$

6.3 Numerical Solution of Acoustic Propagation of Turbulence

While flow turbulence has been of interest for approximately the last 90 years, solution of acoustic noise is of a more recent origin, especially from the 1950s when Lighthill [61] developed the original theory. Thus the initial work was mainly on the basis of Lighthill's theory.

6.3.1 Numerical Solution Through Lighthill's Analogy

Aerodynamically sound is governed by equations that are highly nonlinear, and the technically important flow processes are generally associated with high Reynolds number flows. The computation of acoustic source terms also requires a solution of the time-dependent equations. Initially, the normal approach to acoustic energy calculation is based on a two-step approach consisting of two-step calculations separating noise generation and propagation. In the first step, a stationary mean flow field is calculated by solving the Reynolds-averaged Navier–Stokes equations (RANS) with appropriate turbulence models (for example, with turbulent kinetic energy and turbulent dissipation closure models). However, the drawback of the RANS approach is its limited accuracy in predicting the Reynolds stresses [75] and can provide only time-averaged turbulence parameters, but neither the time history nor the spectra of the fluctuations. Euler equations are then written as an inhomogeneous wave equation,

$$\frac{\partial^2 \rho'}{\partial t^2} - c_\infty^2 \nabla^2 \rho' = \frac{\partial^2 T_{ij}}{\partial x_i \partial x_j}, \tag{6.55}$$

in which $T_{ij} = \rho u_i u_j + \delta_{ij}\left(p' - c_\infty^2 \rho'_{ij}\right) \approx \rho u_i u_j = $ *Lighthill's tensor*, which was written for the fluctuating turbulence term only (fluctuating quadrupole). The full equation, of course, should have terms due to a fluctuating monopole and dipole. Furthermore, we assume that ( ) is constant and the acoustic density function $\rho'/\rho \ll 1$. With all these assumptions the acoustic equation becomes [25]

$$\frac{\partial^2 \rho'}{\partial t^2} - c_\infty^2 \nabla^2 \rho' = \frac{\partial^2 (\rho u_i u_j)}{\partial x_i x_j} + \frac{\partial^2}{\partial x_i x_j}\left[\left(p - \frac{1}{M^2}\rho\right)\delta_{ij}\right], \tag{6.56}$$

where δ_{ij} is the *Kronecker delta*.

The general solution of the inhomogeneous wave Eq. (6.55) is the *Curle equation*, also called the *Lighthill–Curle equation*, which includes also the source at the boundary as

$$\rho'(\mathbf{x},t) = \frac{1}{4\pi c_\infty^2} \int_\Omega \frac{1}{r}\frac{\partial^2 T_{ij}}{\partial y_i \partial y_j}\,d\Omega(\mathbf{y})$$

$$-\frac{1}{4\pi}\int_S \left(\frac{1}{r}\frac{\partial \rho}{\partial n} + \frac{1}{r^2}\frac{\partial \rho}{\partial n}\rho + \frac{1}{c_\infty r}\frac{\partial r}{\partial n}\frac{\partial \rho}{\partial \tau}\right)dS(\mathbf{y}), \tag{6.57}$$

where $\mathbf{x}$ is the position of the observer and $\mathbf{y}$ is the position of the source. It is still written for stationary volume and surface sources, and for moving sources necessary modifications in the equations have to be made.

There are, however, several problems in solving the preceding equation directly, because refraction effects are included in the source term, and in addition, Green's function must be known to obtain an integral formulation. However (6.57) can represent the dynamics of acoustic sources well and the methods such as direct numerical simulation (DNS) are used for this purpose, although the DNS method is very expensive, especially when far-field acoustic computations are required, in terms of the required computational resources. It has been tried only for the simplest geometries [18] due to the wide range of length and time scales present in turbulent flows. Thus, "DNS of high Reynolds number jet flows of practical interest would necessitate tremendous resolution requirements that are far beyond the reach of the capability of even the fastest supercomputers available today" [121].

6.3.2   Kolmogorov Scale

A turbulent flow is composed of eddies of different sizes that define a characteristic velocity scale and time scales (turnover time) dependent on the length scale. The large eddies are mostly unstable and eventually break down into smaller eddies, and the kinetic energy breaks down into smaller eddy energy. The process repeats for ever smaller eddies. In this way the energy is passed down from the largest scales of turbulence until it reaches such a small scale of turbulence that the viscosity of the fluid can effectively dissipate the kinetic energy into *internal energy.*

In his original theory of 1941, *Kolmogorov* postulated that for very high Reynolds numbers, the small-scale turbulence motions are statistically isotropic, that is, no preferential direction could be discerned, and thus this is a measure of the smallest eddies in the flow; typically the tiny eddies are very near the wall region. In general the large scales of turbulence are not isotropic. Kolmogorov's idea was that the statistics of small scales has a uniform character in turbulent flows when the Reynolds number is high.

Thus Kolmogorov introduced a second hypothesis for very high Reynolds numbers – the statistics of small scales are universally and uniquely determined by the kinematic viscosity, $v, m^2/s$, and rate of energy dissipation, $\varepsilon, m^2/s^3$, and from these, by dimensional analysis, a unique length for the scale, called the *Kolmogorov length scale,* η, can be given through the relation

$$\eta = \left(\frac{v^3}{\varepsilon} \right)^{1/4}. \tag{6.58}$$

Once η is divided by the *mesh spacing,* L^Δ, a nondimensional length ratio is obtained, which is what Speziale [107] had used to define his *latency factor* α. This is typically a number between 0 and 1 that defines how much of the local Reynolds stress tensor one will attribute to subgrid (i.e., unresolved) turbulence.

A resolvable scale is one where the wavelength is $\leq 2 * \Delta$ (with Δ being the mesh spacing). The $2 * \Delta$ wavelength is the Nyquist limit of the mesh. Any wavelengths shorter than this cannot possibly be resolved directly and so are modeled (using some form of subgrid treatment). On general 3D meshes, we define Δ as max$[dx, dy, dz]$ because flow structures could exist at any orientation on the mesh.

DNS is based on the direct resolution of the full *Navier–Stokes equations* without any physical assumptions or modes. To get reliable results one must represent all the dynamically active scales of motion in the simulation. This means that the grid spacing Δx and the time step Δt must be fine enough to capture the dynamics of the smallest scales down to the Kolmogorov scale, referred to as η, and the computation domain must be large enough to represent the largest scales.

Resolution requirements referred to a Kolmogorov length scale η at the wall for some incompressible, the wall-bound flows, where x, y, and z refer respectively to the streamwise, wall-normal, and spanwise directions, based on spectral methods, are as follows:

(a) Boundary layer: $\Delta x \cong 15, \Delta y \cong 0.33, \Delta z = 5$;
(b) Homogeneous shear: $\Delta x \cong 8, \Delta y \cong 4, \Delta z \cong 4$;
(c) Isotropic turbulence: $\Delta x = 4, \Delta y = 4.5, \Delta z = 4.5$.

The main constraint in dealing with the resolution in DNS-type computations is that all relevant scales of the flow must be directly captured. The viscous dissipation takes place in a wave band of about $0.1 \leq k\eta \leq 1$, corresponding to a length scale band of 6η to 60η, where η is the *Kolmogarov length scale*. For most realistic applications for shocks, Speziale [107] states that $\eta / \delta_{\text{shock}} \geq 10$.

6.3.3 Smagorinsky model

The approximately constant *Smagorinsky model* was developed by Smagorinsky [103] and is sometimes used to represent unresolved residual motions that manifest themselves in LESs in the form of an unknown residual stress tensor.

Smagorinsky's work was done in connection with the study of the dynamics of the atmosphere's general circulation by numerical integration of a baroclinic primitive equation model for an extended period. In constructing a dynamical model of the atmosphere, one can treat the transient dynamics of the large-scale motions explicitly, or "treat the large scale motions as turbulence which is somehow related to the mean properties of the flow. The latter course has a natural appeal, since it comes from the development of turbulence theory by *Prandtl and von Karman*", and others.

Smagorinsky related the residual stress tensor to the residual strain tensor of the resolved motions through an eddy viscosity. In turn, the eddy viscosity is given by the norm of the resolved strain rate tensor multiplied by the square of a length scale characteristic of unresolved residual motions. The characteristic length scale is set proportional to the primary filter width, wherein the coefficient of proportionality is the *Smagorinsky constant*.

We now discuss the difference between a *resolvable scale* and an *unresolved scale*. A resolvable scale is one in which the wavelength is 2Δ, where Δ is the *Nyquist limit* of the mesh. The *Nyquist frequency*, named after Swedish–American Engineer Harry Nyquist, is a sampling theorem at half the *sampling frequency* of a discrete signal processing system. Any wavelengths shorter than the Nyquist limit cannot possibly be resolved directly and so are modeled (using some form of subgrid treatment). On general 3D meshes, we define Δ as $\max[dx, dy, dz]$ as the flow structures could exist at any orientation on the mesh.

Some researchers have performed LESs of supersonic jet flow and noise using is the *Smagorinsky subgrid scale eddy viscosity model.*

The most popular *subgrid model* is certainly due to Smagorinsky [103] and is obtained by a simple dimensional analysis. Considering that

$$v_t \propto l_o^2, [\mathrm{m^2/s}] \tag{6.59}$$

=*turbulent kinematic viscosity*, we encounter now the problem of evaluating the two characteristic scales l_o and t_o. Assuming that the length scale is representative of the *subgrid modes*, one can write $l_o = C_S\Delta$, where Δ is the grid spacing and C_S is the *Smagorinsky constant*, which was evaluated by *Smagorinsky* as being equal to 0.18 (actually the *Smagorinsky constant* is not constant; the determination of its state is made on a case-to-case basis).

Evaluation of the time scale is somewhat complex and requires new assumptions on the dynamics. We first assume that the local equilibrium hypothesis is satisfied – that is, the production rate of kinetic energy is equal to the transfer rate across the cutoff, which is equal to the dissipation rate by the viscous effect, resulting in an automatic adaptation of the *subgrid scale* to the resolved ones (that is, the information propagates at an infinite speed along the spectrum). The value of the *Smagorinsky constant* is now a different one and depends a priori on the energy spectrum; a lower value of typically 0.1 should be used for a channel flow. The characteristic time scale of the subgrid modes is then equal to that of the resolved scales, which is assumed to be the turnover time defined as

$$T_0 = \frac{1}{\sqrt{2\bar{S}_{ij}\bar{S}_{ij}}} = t_o, \tag{6.60}$$

leading to the *turbulence kinematic viscosity*

$$v_t = (C_S\Delta)^2 \sqrt{2\bar{S}_{ij}\bar{S}_{ij}}, \tag{6.61}$$

where Δ is the LES cutoff length $[m]$ and S_{ij} $[s^{-1}]$ is the *mean stress tensor* given by

$$\bar{S}_{ij} = \left(\frac{\partial \bar{u}_i}{\partial x_j} + \frac{\partial \bar{u}_j}{\partial x_i}\right) - \frac{2}{3}\frac{\partial \bar{u}_k}{\partial x_k}\delta_{ij}. \tag{6.62}$$

Uzun et al. [122] have studied the sensitivity to the *Smargorinsky constant* in turbulent jet simulations. "With the recent improvements in the processing speed of computers, the application of *direct numerical simulation* (DNS) and *large eddy simulation* (LES) to jet noise prediction methodologies is becoming more feasible. However, due to the wide range of length scales and time scales present in turbulent flows, DNS is still restricted to *low Reynolds number flows* in relatively simple geometries. LES, with lower computational cost, is an attractive alternative to DNS."

In an LES, the flow field is decomposed into a large-scale or resolved-scale component and a small-scale or subgrid-scale component by filtering the entire domain using a grid filter function G with filter width Δ. The filtering operation removes the small-scale or subgrid-scale (SGS) turbulence from the *Navier–Stokes equations*, and the resulting governing equations are then solved directly for the large-scale turbulent motions, while the effect of the SGSs is computed using a subgrid-scale model, for example the classical *Smagorinsky model.*

The model sensitivity is determined by applying it to the LES of turbulent jets for two simulations with model constants 0.018 and 0.019 and the results of *turbulent decay coefficients* were compared

The potential core of the jet breaks up around $x = 14r_o$, and after transition to turbulence, a linear growth, consistent with the experimental observation, is obtained. Half of the inverse of this line's shape, which is equal to the jet decay coefficient, was found to be approximately 5.0 for a simulation with 0.018, while it was 5.76 for a simulation with 0.019, when the experimental values of the jet decay coefficient for the jet spreading were found to be between 5.4 and 6.1. Hence the jet decay coefficient is directly affected by the *Smagorinsky model constant.* For a 5.56% increase in the model constant, the jet decay coefficient increases by 15.20%.

Another comparison is made of the *normalized shear stress* in the jet mixing region and compared with the experimental results, and a similar effect of the *Smagorinsky constant* is obtained.

6.3.4 Kirchoff Surface Formulation

For a body moving in air, the main disturbance is in the vicinity of the body, whereas at a large distance the disturbance may be considered in a cylindrical coordinate system. A cylindrical *Kirchoff surface* S is assumed to enclose all the nonlinear effects and sound sources. Outside the surface the flow may be considered to be linear and be governed by the convective wave equation (for subsonic flow)

$$\nabla^2 \varphi - \frac{1}{c_\infty^2}\left(u_\infty \frac{\partial}{\partial x} + \frac{\partial^2}{\partial t^2} \right)\varphi = 0, \tag{6.63}$$

where $\varphi = -\nabla V$ is the well-known velocity potential. If the background velocity u_∞ is zero, the preceding equation reduces to the simple wave equation.

"Surface integral methods can efficiently and accurately predict generated noise provided the control surface surrounds the entire source region, where the linear wave equation is valid. However, for jet noise prediction, shear mean flow exists outside the control surface that causes refraction" [87, 88]. Their proposed methodology shows that the prediction of the far-field sound pressure level for the acoustic field within the *cone of silence* is consistent with the measured data. Additional nonlinearities, for example, can be added outside the control *Kirchoff surface.*

In the classical Kirchoff formulation for stationary ambient fluid, the far-field sound pressure is given by a surface integral involving the pressure and its normal derivatives at the surface. Now the *Green function* as a solution of the equation is

$$\nabla^2 G - \frac{1}{c_\infty^2}\left(u_\infty\frac{\partial}{\partial x} + \frac{\partial^2}{\partial t^2}\right)^2 G = \delta(x-x',y-y',z-z',t-t'), \qquad (6.64)$$

where c_∞ is the free stream speed of sound, u_∞ is the free stream velocity, and δ is the Dirac function. The source location is (x',y',z',t') at retarded time t' and (x,y,z,t) is the observer's location; the time delay is given by $(\tau = t - t' = r/c)$ between the emission of the sound and detection.

According to the *small perturbation theory* in aerodynamics, there is a similarity in space and time for which, in subsonic flow, the distance between the observer and the surface point in *Prandtl–Glauert coordinates* is

$$x_o = x, y_o = y\beta, z_o = z\beta, r_o = \sqrt{(x-x')^2 + \beta^2[(y-y')^2 + (z-z')^2]}$$

$$\tau = \frac{[r_o - M_\infty(x-x')]}{c_\infty\beta^2}, \beta = \sqrt{1-M_\infty^2},$$

where subscript o denotes the Prandtl–Glauert base value ($= 0$ for subsonic flow and $= \sqrt{2}$ is for supersonic flow, as explained below), M_∞ is the free stream Mach number, and τ is the time delay (r/c), but $\tau' = t - \tau$ is the retarded time.

The convective wave equation in the subsonic case is given by the solution

$$M_\infty G = -\frac{\delta}{\tau'}, \qquad (6.65)$$

where G is the *Green's function* and the *potential function solution* is

$$4\pi\varphi(x,t) = \int_{S_o}\left[\frac{\varphi}{r_o^2}\frac{\partial r_o}{\partial n_o} - \frac{1}{r_o}\frac{\partial\varphi}{\partial n_o} + \frac{1}{cr_o(1-M_o^2)}\frac{\partial\varphi}{\partial\tau}\frac{\partial r_o}{\partial n_o} - M_\infty\frac{\partial x_o}{\partial n_o}\right]_\tau dS_o.$$

$$(6.66)$$

Equation (6.66) is an integral representation of φ at points exterior to S in terms of information prescribed on control surface S and can be used to obtain a solution at an external arbitrary point if the solution is known at S, but inside the surface $\varphi = 0$.

For a supersonically moving surface the solution of *Green's function* remains (6.65) and the governing equation remains (6.63). However, for a supersonically moving surface the time delay τ is not uniquely defined. The time delay is now

$$\tau^\pm = \frac{[\pm r_o - M_\infty(x-x')]}{c_\infty\beta^2}, \beta = \sqrt{M_\infty^2 - 1} \qquad (6.67)$$

and the *potential function solution* becomes

$$4\pi\varphi(x,t) = \int_{S_o}\left[\frac{\varphi}{r_o^2}\frac{\partial r_o}{\partial n_o} - \frac{1}{r_o}\frac{\partial \varphi}{\partial n_o} + \frac{1}{cr_o\beta^2}\frac{\partial \varphi}{\partial \tau}\pm\frac{\partial r_o}{\partial n_o} - M_\infty\frac{\partial x_o}{\partial n_o}\right]_\tau \mathrm{d}S_o. \quad (6.68)$$

Further, we can write the solution in the frequency domain. For this let

$$\varphi(\mathbf{x},t) = \mathrm{REAL}\cap\varphi(\mathbf{x},t). \quad (6.69)$$

Substituting (6.69) into (6.66) and accounting for the retarded time, we obtain the *Kirchhoff's formulation* in the *frequency domain* for a uniformly moving surface for a 2D problem. As pointed out by [66] the method has not been extended for the 3D case.

6.3.5   Boundary Conditions

Attempts to specify nonreflecting boundary conditions for realistic flow problems have followed three main paths: (a) characteristic-based boundary conditions, (b) linearized asymptotic boundary conditions, and (c) absorbing boundary conditions. The first will be discussed now, especially for steady or quasisteady boundary conditions.

Given are the 2D Euler equations in nonconservative form as

$$\rho_t + u\rho_x + v\rho_y + \rho\,(u_x + v_y) = 0,$$

$$u_t + uu_x + vu_y + (1/\rho)p_x = 0,$$

$$v_t + uv_x + vv_y + (1/\rho)p_y = 0,$$

$$p_t + up_x + vp_y + \rho c^2\,(u_x + v_y) = 0.$$

These can be written in vector form as

$$\mathbf{Q}_t + \mathbf{A}\mathbf{Q}_x + \mathbf{B}\mathbf{Q}_y = 0, \quad (6.70)$$

where

$$\mathbf{Q} = (\rho,u,v,p), \quad (6.71)$$

$$\mathbf{A} = \begin{bmatrix} u & \rho & 0 & 0 \\ 0 & u & 0 & 1/\rho \\ 0 & 0 & u & 0 \\ 0 & \gamma p & 0 & u \end{bmatrix}, \quad (6.72)$$

$$\mathbf{B} = \begin{bmatrix} v & 0 & \rho & 0 \\ 0 & v & 0 & 0 \\ 0 & 0 & v & 1/\rho \\ 0 & 0 & \gamma p & v \end{bmatrix}. \quad (6.73)$$

Here $\mathbf{Q}$ is a primitive variable vector in terms of ρ, u, v, p. These are, of course, not the only independent variables, and we will note later the consequence of writing at least two of these in terms of vorticity χ and entropy s.

We define vorticity by

$$\chi = u_y - v_x. \tag{6.74}$$

By forming a y-derivative of the second row of matrix Eq. (6.70) and an x-derivative of the third row of the equation, under the condition that the variables before the derivatives are kept momentarily constant, and subtracting from one another, we get the vorticity equation

$$\chi_t + u\chi_x + v\chi_y = 0. \tag{6.75}$$

Now we write the energy equation from the *first law of thermodynamics* as

$$T\mathrm{d}s = c_p\mathrm{d}T - \frac{1}{\rho}\mathrm{d}p, \tag{6.76}$$

from which, after some manipulation, we get

$$\mathrm{d}s = c_p\frac{\mathrm{d}T}{T} - R\frac{\mathrm{d}p}{p} = c_p\left(\frac{\mathrm{d}p}{p} - \frac{\mathrm{d}\rho}{\rho}\right) = (c_p - R)\frac{d}{p} - c_p\frac{\mathrm{d}\rho}{\rho}, \tag{6.77}$$

and hence

$$\frac{\mathrm{d}s}{c_p} = \frac{1}{\gamma}\frac{\mathrm{d}p}{p} - \frac{\mathrm{d}\rho}{\rho}. \tag{6.78}$$

Therefore, from the first and fourth rows of (6.70) it can be shown easily that

$$s_t + us_x + vs_y = 0. \tag{6.79}$$

We will now examine the implication of (6.70). For this the $\mathbf{A}$ matrix in (6.73) is further manipulated to get

$$\mathbf{A} = \mathbf{T}^{-1}\boldsymbol{\Lambda}\mathbf{T}, \tag{6.80}$$

where

$$\boldsymbol{\Lambda} = \begin{bmatrix} u & 0 & 0 & 0 \\ 0 & u+c & 0 & 0 \\ 0 & 0 & u & 0 \\ 0 & 0 & 0 & u-c \end{bmatrix}, \tag{6.81}$$

$$\mathbf{T} = \begin{bmatrix} c^2 & 0 & 0 & -1 \\ 0 & \rho c & 0 & 1 \\ 0 & 0 & 1 & 0 \\ 0 & -\rho c & 0 & 1 \end{bmatrix}, \tag{6.82}$$

$$\mathbf{T}^{-1} = \begin{bmatrix} 1/c^2 & 1/(2c^2) & 0 & 1/(2c^2) \\ 0 & 1/(2\rho c) & 0 & -1/(2\rho c) \\ 0 & 0 & 1 & 0 \\ 0 & 1/2 & 0 & 1/2 \end{bmatrix}, \tag{6.83}$$

and we write the vector equation as

$$\mathbf{Q}_t + \mathbf{Q}_x = -\mathbf{B}\mathbf{Q}_y = \mathbf{S}. \tag{6.84}$$

Multiplying the preceding equation by $\mathbf{T}$ we get

$$\mathbf{T}\mathbf{Q}_t + \Lambda \mathbf{T}\mathbf{Q}_x = -\mathbf{T}\mathbf{B}\mathbf{Q}_y = \mathbf{T}\mathbf{S} = \mathbf{S}'. \tag{6.85}$$

Noting that

$$\mathbf{T}\mathbf{Q} = \mathbf{Q}' = \rho c^2 - \rho\rho cu + pv - \rho cu + p, \tag{6.86}$$

the matrix equation becomes

$$\mathbf{Q}'_t + \Lambda \mathbf{Q}'_x = \mathbf{S}'. \tag{6.87}$$

The equation can be written in characteristic form as

$$\frac{d\mathbf{Q}'}{dC} \equiv \frac{\partial \mathbf{Q}'}{\partial t} + \Lambda \frac{\partial \mathbf{Q}'}{\partial x} = \mathbf{S}', C = x \pm ct, \tag{6.88}$$

and it can be shown that $\mathbf{Q}'$ is invariable in the characteristic directions for which the propagating speed is given by the diagonal matrix, Λ. For the second and fourth rows of the characteristic equations the propagating speeds are $(u \pm c)$, while the characteristic variables are $(p \pm \rho cu)$, but from the first and third rows it is seen that the characteristic speed is u. It was shown also in (6.75) and (6.79) that propagation of vorticity and entropy is also along the u-direction.

Out of the four eigenvalues (first and third combined as u, $u + c$, and $u - c$) and for a flow left to right $(u > 0)$ it is obvious that there are three positive eigenvalues for the outflow boundaries and for a subsonic flow there is one negative eigenvalue $(u - c)$ as inflow boundary condition. It has been pointed out by Thompson [115] that "at a point on (say) the x-boundary, some number of characteristic velocities describe outgoing waves, while some of them describe incoming waves. The behavior of the outgoing waves is completely determined by data combined within and on the boundary, while the behavior of the incoming waves is specified by data external to and on the boundary. The number of boundary conditions which must be specified at a point on the boundary is equal to the number of incoming waves at that point".

The preceding equation is now discretized for basically a steady problem as follows: Let, at time t_n, $\mathbf{Q}'$ be known on the boundary point $(3')$ and on the next point $(1')$. On the other hand the value of $\mathbf{Q}'$ is not known on the boundary (Q'_3), but it is known at the next point (Q'_1). Let there be a point $2'$ at time t_n on $\mathbf{x}$ on which the characteristic from point 3 falls. Obviously, $x_1 = x_{1'}$ and $x_3 = x_{3'}$. By interpolation we get

$$x_{2'}x_3 - \frac{\Delta}{2}(\Lambda_{2'} + \Lambda_{3'}) \text{ and also } \mathbf{Q}'_{2'} = \mathbf{Q}'_{1'} + (Q'_{3'} - Q'_{1'})\frac{x_{2'} - x_1}{x - x_1}. \tag{6.89}$$

Further,

$$\mathbf{Q}_3' = \mathbf{Q}_2' + \frac{\Delta t}{2}(\mathbf{S}_2' + \mathbf{S}_3'). \tag{6.90}$$

With initial guess values of $\mathbf{Q}_3' = \mathbf{Q}_1'$, $\Lambda_3 = \Lambda_1$, and $\mathbf{S}_3' = \mathbf{S}_1'$, the solution of $\mathbf{Q}_1$ is obtained iteratively for the incoming characteristic and the outgoing characteristic is set equal to zero.

While imposing a constant prescribed pressure at a subsonic exit condition, since both $(u + c)$ and $(u - c)$ characteristic speeds are applicable, one has to show actually that the perturbation waves are to be reflected at the (subsonic) exit boundary and thus the change in pressure $dp = 0$ is eliminated from equations in both characteristic directions to obtain $d\rho$ and du.

One could do a similar analysis if the boundary is not perpendicular to the x-coordinate. In such a case the original set of differential equations (a–d) in the (x, y, t) coordinate directions are converted into (ξ, η, t) coordinates, where (ξ, η) is again a 2D orthogonal coordinate system but the ξ-axis is normal to the boundary.

6.4   Various Computational Aeroacoustic Methods

6.4.1   Direct Numerical Simulation and Direct Noise Computation Methods

The straightforward method for broadband noise is to solve the compressible Navier–Stokes equations using direct numerical simulations. Resolving all scales of a turbulent flow in a DNS requires that the smallest turbulent scales be solved on extremely fine grids with increasing Reynolds numbers. For 3D flows, the requirement is for computation times that scale with the third power of the Reynolds number. Thus for most applications, the high Reynolds number is a constraint. In addition, a huge computation domain has to be chosen to simulate the far field. Nevertheless, direct numerical calculations have been performed for simple geometries, for example the 2D driven cavity flow, and also for turbulent boundary layer flow by Spalart [104].

The classical fourth-order Runge–Kutta scheme has been used in CAA for the calculation of near-field sources, for example jet flows by Freund [37] and in a mixing layer by *Colonius et al.* [33].

The first simulations of free shear flows to compute their radiated noise from the unsteady compressible flow equations were performed using DNS of bidirectional and axisymmetric groups by the Stanford group of Colonius et al. [33] and Mitchell et al. [77], respectively, of the sound generated by vortex pairing in an axisymmetric jet. They showed the feasibility of *direct noise computation* (DNC) by directly calculating the noise generated by vortex pairing in mixing layers. Afterward, the DNC was also applied with DNS for 3D mixing layers. A steady supply of acoustic sources in a turbulent jet at Mach 0.9 and Reynolds number 3,600 were done in a DNC study by Freund [37].

To simulate the jet, the compressible flow equations were formulated in cylindrical coordinates and solved without modeling approximations, with additional details concerning boundary conditions used in a spatially developing computation [37]. Sixth-order compact finite-difference schemes were used to compute derivatives in the axial and radial directions, and Fourier spectral methods were used in the azimuthal direction. The solution was advanced in time with a fourth-order Runge–Kutta algorithm.

The jet exit conditions to estimate *approximate* nozzle conditions for use in the computations were given for the mean flow by assuming a rounded top-hat profile that is often used in jets:

$$\frac{\bar{v}_r}{U_{\text{jet}}} = \frac{1}{2}\left[1 - \tanh\left[b(\theta,t)\left(\frac{r}{r_o} - \frac{r_o}{r}\right)\right]\right], \qquad (6.91)$$

where r_o is the jet nozzle radius, U_{jet} is the exit velocity, $\bar{v}$ is believed to be the average perturbation velocity, and b is a thickness parameter in a *random-walk fashion* about a mean value of $\bar{b} = 12.5$, which induced fluctuations (typically smaller than 1%) in the shear-layer thickness near the nozzle. This particular form of excitation was chosen for this application "because it is approximately irrotational which minimized spurious noise generation. The results, except for a weak dependence of the potential core length on the initial amplitude of the initial amplitude of the perturbations, were not found strongly on the particular nature of the excitation [37]." Complete documentation of the randomization technique was provided by *Freund* in one of his papers.

"The physical (nonboundary zone) portion of the computational domain extended $30r_o$ in the streamwise direction and $8r_o$ in the radial direction. The computational mesh consisted of $640 \times 250 \times 160$ points in the axial, radial, and azimuthal directions, respectively. It was compressed radially around $r = r_o$ and axially around $x = 15r_o$. Approximately 50,000 time steps were computed after the flow achieved a statistically stationary state [37]." Comparison of computed results with experimental data was found to be excellent.

Introducing natural disturbances, no spurious disturbance is generated in the vicinity of the inlet plane. For the avoidance of such difficulties, two approaches have been used successfully used to date [12]. The first method, which attempts to construct spurious disturbance-free or nearly disturbance-free is due to Bogey et al. [27] and *Uzun* [119, 121] by introducing disturbances of the form

$$u'_x = f_1(\theta)g_1(t)h_1(x,r),$$

$$u'_r = f_2(\theta)g_1(t)h_2(x,r),$$

$$u_\theta = 0$$

where the functions h_1 and h_2 are chosen such that the functions $\nabla \bullet \mathbf{u}' = 0$. This technique works well for cold low and moderate Mach number jets where the density is nearly uniform. Similarly, one may vary the inlet momentum thickness as a function of time and azimuthal location.

The second approach uses the nonradiating eigenfunction solutions as a linearized spatial instability problem associated with the inlet mean profile as the disturbance profile. Assuming a modal decomposition of a disturbance flow variable,

$$q'(x,r,\theta,t) = \hat{q}(r)\exp^{i(\omega t - kx - n\theta)} \qquad (6.92)$$

for integer n for an axisymmetric mean flow is governed by the *cylindrical Rayleigh equation*

$$\frac{d^2\hat{p}}{dr^2} + \left[\frac{1}{r} - \frac{1}{\bar{\rho}}\frac{d\bar{\rho}}{dr} + \frac{2k}{(\omega - k\bar{u})}\frac{d\bar{u}}{dr}\right]\frac{d\hat{p}}{dr} + \left[\frac{(\omega - k\bar{u})^2}{(\bar{a})^2} - \frac{n^2}{r^2} - k^2\right]\hat{p} = 0 \qquad (6.93)$$

subject to the boundary conditions

$$|\hat{p}| \leq \infty \text{ as } r \to 0,$$

$$\hat{p} \to 0 \text{ as } r \to \infty.$$

The spatial stability problem is characterized by fixing ω to be real and solving for the *complex eigenvalue k arc*. The remaining flow variables are given as linear combinations $\bar{p}$ and their radial derivative [12].

Using the pressure fluctuations on the cell, the preceding equation can be solved in partially transformed form as a *Bessel equation*,

$$\left(\frac{d^2}{dr^2} + \frac{1}{r}\frac{d}{dr} + \frac{\omega^2}{a_\infty^2} - k^2 - \frac{n^2}{R^2}\right)\hat{p} = 0, \qquad (6.94)$$

that is based on the direct resolution of the field unsteady Navier–Stokes equations without any physical assumptions or models. To get all the dynamic scales of motion in the simulation, the grid spacing Δx and the time step Δt must be fine enough to capture the dynamics of the smallest scales of the flow down to the *Kolmogorov scale* and the computational domain must be large enough to represent the large scales. These criteria lead to a high computational cost, which is responsible for the fact that DNS is nowadays used only for turbulent jets [38] and in a mixing layer by Colonius et al. [33].

The predicted acoustic far field was found to agree with Lighthill's acoustic analogy. The same year, Freund [37] performed 3D DNS of a randomly forced round jet at $Re = 3.6 \times 10^3$ and $M = 0.9$ using 25 million grid points. Predicting the far-field acoustic pressure by solving a wave equation within the near-field pressure data from DNS, they also obtained [33, 37] good agreement with the experimental data.

Because DNS requires very fine grid resolution and therefore very high CPU time and memory, currently it is only affordable for very low Reynolds numbers and simple geometries. A good compromise is LES, which resolves a fairly large range of frequencies and models only a small part of the flow.

6.4.2   *Linearized Euler Equations*

Because the amplitude of sound waves is much smaller than any other flow variables, the equations can be linearized. Linearization is done by defining each variable as a sum of one average part and a perturbed part; when two such variables are multiplied by each other and the product of the two perturbed variables is neglected, then the linearized equation is obtained. The LEEs are a natural extension to Lighthill's analogy in CAA and provide accurate numerical solutions by dealing only with perturbation and taking into account the refraction effects of sound waves induced by the mean flow [18]. The approach is to add a source term expression on the right-hand side of the differential equation from a synthesized turbulence field. However, some difficulties remain concerning the axisymmetric calculation of noise propagation since acoustic sources are completely correlated in the azimuthal direction in certain formulations. In 2D studies the starting point by the authors is an LEE around a stationary mean flow

$$\frac{\partial \mathbf{U}}{\partial t} + \frac{\partial \mathbf{E}}{\partial x} + \frac{\partial \mathbf{F}}{\partial y} + \mathbf{H} = \mathbf{S}, \tag{6.95}$$

where

$$\mathbf{U} = \left(\rho', \overline{\rho}u', \overline{\rho}v', p'\right),$$
$$\mathbf{E} = \left(\rho'\overline{u} + \overline{\rho}u', \overline{\rho u}u' + p', \overline{\rho u}v'\overline{u}p' + \gamma\overline{p}u'\right),$$
$$\mathbf{F} = \left(\rho'\overline{v} + \overline{\rho}v', \overline{\rho v}u', \overline{\rho v}v' + p', \overline{v}p' + \gamma\overline{p}v'\right),$$
$$\mathbf{H} = \left(0, (\overline{\rho}u' + \rho'\overline{u})\overline{u}_x, (\overline{\rho}v' + \rho'\overline{v})\overline{v}_y, 0\right).$$

Here $\mathbf{E}$ and $\mathbf{F}$ are the 2D flux vectors, and the vector $\mathbf{H}$ contains terms related to the gradients of the mean flow, which are equal to zero if the mean flow is uniform. The source term $\mathbf{S}$ is given by [19]

$$\left[0, S_1 = S_1^f - \overline{S_1^f}, S_2 = S_2^f - \overline{S_2^f}, 0\right], \tag{6.96}$$

where the source term must be determined from the fluctuating velocity field and not from the instantaneous velocity field as

$$S_i^t = \frac{\partial \rho u_i u_j}{\partial x_j} - \frac{\overline{\partial \rho u_i' u_j'}}{\partial x_j}$$

$$= \frac{\partial \rho \overline{u}_i u_j'}{\partial x_i} + \frac{\partial \rho u_i' \overline{u}_j}{\partial x_j} + \frac{\partial \rho u_i' u_j'}{\partial x_j} \frac{\overline{\partial \rho u_i' u_j'}}{\partial x_j}, \tag{6.97}$$

where the third and fourth terms on the right-hand side is $S_i^f - \overline{S_i^f}$.

Thus the vector $\mathbf{S}$ represents possible unsteady sources in the flow. For a single fluctuating monopole source this can, for example, be defined by writing

$$S(x,y,t) = f(x,y)\sin(\omega t)(1,0,0,1),$$

$$f(x,y) = A\exp^{-\alpha[(x-x_o)^2+(y-y_o)^2]},$$

where A is the source amplitude and ω is the angular radian frequency. Similarly, a dipole may be defined as a divergence of a source term, $\mathbf{S} = -\nabla\cdot\mathbf{F}$, where, for example,

$$F_i = \varepsilon\cos\left[(\pi/10)x\right]\exp^{-\alpha y^2}\sin(\omega t), \tag{6.98}$$

and for a single quadrupole the quadrupole is written as

$$T_{ij} = \begin{bmatrix} -\cos[(\pi/20)x]\exp^{-\alpha y^2} & 0 \\ 0 & -\cos[(\pi/20)y]\exp^{-\alpha x^2}] \end{bmatrix}\frac{20A}{\pi}\sin(\omega t). \tag{6.99}$$

The solution of ρ' is then given in terms of *Green's function*.

After a sound source is predicted, the obvious method should be to compute its propagation by extending the computational domain all the way up to the observer. However, if the objective is to compute the far-field noise, this direct approach will be prohibitive in terms of the computer resources without much benefit. In addition, using nonlinear flow equations there can be an accumulation of errors, especially because acoustic fluctuations are much smaller than aerodynamic fluctuations. It is, therefore, necessary to split the computation into two separate domains, one describing the nonlinear propagation of sound and the other describing the linear propagation of acoustic signals.

In order to solve the general field problem, we seek the solution of the preceding equation for a point source of unit strength in the form of a *delta function* at $t = t_o$ and $\mathbf{r} = \mathbf{r}_o$, and we seek the solution of the equation with *Green's function* as

$$\frac{1}{c^2}\frac{\partial^2 G(\mathbf{r},t\mid\mathbf{r}_o,t_o)}{\partial t^2} - \nabla^2 G(\mathbf{r},t\mid\mathbf{r}_o,t_o) = 4\pi\delta(\mathbf{r}-\mathbf{r}_o)\delta(t-t_o). \tag{6.100}$$

As an initial condition it may be considered reasonable to assume that both G and $\partial G/\partial t$ are zero when $t < t_o$. In addition, we consider generally an outward-propagating wave to infinity, where G must vanish. Therefore, we seek a solution (Morse and Feshbach [5]),

$$G(r\mid r_o) = \exp^{ikR}/R, k = (\omega/c)^2, \tag{6.101}$$

where

$$R = \mid r-r_o\mid = \sqrt{(x-x_o)^2+(y-y_o)^2+(z-z_o)^2}. \tag{6.102}$$

Thus we divide the computational domain into two parts: (1) the inner domain with a boundary where the Kirchhoff surface is constructed using existing grid lines outside of the sound source zone but well within the Navier–Stokes computational domain and (2) the Kirchhoff method, by a surface integral method based on the formulation of Green's equation for the linear wave equation, is used to predict the far-field noise.

6.4.3 Limited Numerical Scales Concepts

The limited-numerical scales (LNS) approach was inspired by an earlier suggestion of Speziale [107]. In the *LNS approach*, the definition of a *latency factor* α is introduced, in which the *stress tensor* $\overline{u_i'u_j'}$ is damped via

$$\alpha = \frac{\overline{u_i'u_j'}}{\overline{u_i'u_j'}^M} = \left[1 - \exp^{-\beta L_i^\Delta / L^k}\right]^n \tag{6.103}$$

$$L_i^\Delta = 2\max\left[\Delta x, \Delta y, \Delta z, \sqrt{(u_i^2)}/\Delta t\right], \tag{6.104}$$

where β and n are some unspecified parameters, L^k is some representative *mesh spacing*, and $L^k = v^{3/4}/\varepsilon^{1/4}$ is the *Kolmogorov length scale*. Superscript M denotes values computed using the conventional *Reynolds-averaged Navier–Stokes (RANS) equations*.

Now the *kinematic turbulence viscosity model* [m²/s] from the *conventional Smagorinsky model* [103] $[S_{kl}^*]$ is given as

$$\mu_t = C_s \left(L_i^\Delta\right)^2 \sqrt{S_{kl}^* S_{kl}^*/2}, \tag{6.105}$$

where C_s is the *Smagorinsky constant*, L^Δ defines a filter length, which allows between the unresolvable and resolvable scales of motion, $S_{ij}^*[\mathrm{s}^{-1}]$ is the *mean strain tensor*. S is the nondimensional strain magnitude, $\omega_{ij}[\mathrm{s}^{-1}]$ is the vorticity and ω is the dimensionless vorticity. The *Smagorinsky constant* is not actually a universal constant and has a value here of 0.05.

A *reasonable scale* is one where the wavelength $\leq 2\Delta$, with Δ being the mesh spacing and 2Δ the *Nyquist limit* of the mesh.

The governing equations, in addition to standard continuity and momentum equations, are slightly modified from the standard (k, ε) model in (6.46) and (6.47) as follows:

k equation:

$$\frac{D\bar{\rho}\bar{k}}{Dt} = \frac{\partial}{\partial x_i}\left[\left(\mu + \frac{\mu_t}{\sigma_k}\right)\frac{\partial \bar{k}}{\partial x_i}\right] + \rho\bar{\varepsilon}_m + \left(\frac{\partial \bar{u}_i}{\partial x_j} + \frac{\partial \bar{u}_j}{\partial x_i}\right)\frac{\partial \bar{u}_i}{\partial x_j} - \rho\varepsilon$$

$$= \frac{\partial}{\partial x_i}\left[\left(\mu + \frac{\mu_t}{\sigma_k}\right)\frac{\partial \bar{k}}{\partial x_i}\right] + P_k - \bar{\rho}\bar{\varepsilon}; \tag{6.106}$$

ε equation:

$$\frac{D\bar{\rho}\bar{\varepsilon}}{Dt} = \frac{\partial}{\partial x_k}\left[\left(\mu + \frac{\mu_t}{\sigma_\varepsilon}\right)\frac{\partial\bar{\varepsilon}}{\partial x_k}\right] + c_{\varepsilon 1}\frac{\varepsilon}{k}\varepsilon_m\left(\frac{\partial\bar{u}_i}{\partial x_j} + \frac{\partial\bar{u}_j}{x_i}\right)\frac{\partial\bar{u}_i}{\partial x_j} - c_{\varepsilon 2}\frac{\varepsilon^2}{k}$$

$$= \frac{\partial}{\partial x_k}\left[\left(\mu + \frac{\mu_t}{\sigma_\varepsilon}\right)\frac{\partial\bar{\varepsilon}}{\partial x_k}\right] + (c_{\varepsilon 1}P_k - c_{\varepsilon 2}\bar{\rho}\bar{\varepsilon} + E)\,T_t^{-1}, \qquad (6.107)$$

where the *turbulent energy production rate* is $P_k[\mathrm{kgm^{-1}s^{-3}}]$ is the power $[\mathrm{kgm^2 s^{-3}}]$ per unit of volume $[\mathrm{m^{-3}}]$, and is given by the relation

$$P_k = -\rho\overline{u_i'' u_j''}\frac{\partial\bar{u}_i}{\partial x_j}. \qquad (6.108)$$

Equation (6.107) is slightly different from (6.47) except for T_t^{-1}. This is because it is a *realizable time scale* – one that reads a finite value at the limit of low Reynolds numbers, and it is simply set equal to ε/k, which leads to a singularity at the wall, where $k \to 0$.

6.4.4 Detached Eddy Simulation

The *detached eddy simulation* (DES) was the most widely used *hybrid method* in the 2000–2004 time frame thanks to clear objectives, open publication, resources for detailed testing by Strelets [109], commonality with a known RANS model, and a simple and complete formulation by *Spalart et al.* [105, 106]. For this, DES takes the classical view that the filter width is best tied to the grid spacing in order to make the best use of the available resolution. The length scale injected into the model is

$$\tilde{d} \equiv \min(d, C_{\mathrm{DES}}\Delta), \qquad (6.109)$$

where d is the traditional wall distance. In the entire boundary layer $\tilde{d} = d$ is used, whereas in separated regions, given an adequate grid, $\tilde{d} = C_{\mathrm{DES}}\Delta$ is used, which has many similarities with the *Smagorinsky model*, certainly its scaling with grid spacing and *dissipation rate*.

The objectives of DES overlap with those of [107], but the focus on treating the entire boundary layer with RANS appears to represent a clear difference. In addition, DES practice has benefited from the routine control of the transition in the *Spalart et al.* [105] model, for which there is no better demonstration than the circular cylinder case (Travin et al. [117]) by representing the "most qualitative features of the flow in addition to nontrivial qualitative features such as three dimensional chaos and intense modulation of vortex shedding" [12].

As of 2004, DES enjoyed a large user base and a publication stream. "The principal cause for concern had been in cases in which the grid is progressively refined until $\tilde{d} = C_{\mathrm{DES}}\Delta$ intrudes inside the boundary layer. The result is a weakened eddy viscosity but one that is not weak enough to allow LES eddies to form; as a result, the separation line moves too far forward. …DES is now appearing in vendor *computational fluid dynamics* (CFD) codes, in which it has been demonstrated with very uneven levels of understanding" [12].

"Of the *hybrid methods* described, the DES approach has the widest experience base. The simplicity of the DES formulation is clearly appealing and is likely to be make the first method of choice for *hybrid* RANS-LES implementation in new or existing CFD codes" [12]. The differences between the RANS-LES formulation and the LNS method "are expected to be small when used as explicitly intended by the DES."

"The DES approach can still be considered a *zonal method* because the two domains (LES-RANS) are fully determined by the grid topology and segmentation is fully independent of the flow solutions. In DES, attached flow regions (*attached eddies*) are distinguished from the separated flow regions (*detached eddies*). The former are properly predicted based on RANS with statistical turbulence models, whereas the latter, including large-scale unsteady vortices, are computed more reasonably by LES. Thus, DES could be represented as a natural hybrid method combining RANS and LES" [12].

The 2001 American Institute of Aeronautics and Astronautics meeting paper by Strelets [109] attempted to provide "a comprehensive description of the state-of-the-art in the area of the *Detached Eddy Simulation* of massively separated turbulent flows. It soundly combines fine-tuned Reynolds-averaged Navier–Stokes (RANS) technology in the attached boundary layers and the power of the Large-Eddy Simulation (LES) in the separated regions. It is essentially a 3D unsteady approach using a single turbulence model, which functions as an *SGS model* in the regions where the grid density is fine enough for an LES, and as a RANS model in regions, where it is not. The 'fine enough' for an LES is that one whose maximum 3D grid step, Δ, is much smaller than the flow turbulence length scale, δ_t (this is an integral length scale of the turbulence, much larger than the *Kolmogorov length scale*, of course). The Smagorinsky (SGS) function or LES mode prevails where the grid spacing in all directions is much smaller than the thickness of the *turbulence shear layer*. The model senses the grid density and adjusts itself to a lower level of mixing, relative to the RANS mode, and, as a result, unlocks the large-scale instabilities of the flow and lets the energy cascade close to the grid spacing. In other regions (primary attached boundary layers), the model is in RANS mode. The approach is nonzonal, that is, there is a single velocity and pressure field and no issue of smoothness between the regions. The computer-cost outcome is favorable enough that challenging flows at high Reynolds numbers can be treated quite successfully on the latest personal computers. The credibility of the approach is supported by a set of numerical examples of its application: NACA 0012 airfoil at high (up to 90°) angles of attack, circular cylinder with laminar and turbulent separation, backward-facing step, triangular cylinder in a plane channel, raised airport runway, and a model of a landing gear truck. The DES predictions are compared with experimental data and with RANS solutions [12]."

The driving length scale of the DES is the distance to the closest wall, d_w, which is substituted everywhere with a new DES length scale, $\bar{l}$; it is defined as

$$\bar{l} = \min(d_w, C_{\mathrm{DES}}\Delta), \tag{6.110}$$

where C_{DES} is only adjustable new model constant.

On the other hand, in the $\mathrm{M_{SST}}$ formulation, in which two turbulence equations model, $(k - \varepsilon)$ or $(k - \omega)$, is used, the dissipative term of the k-transport equation for the $(k - \omega)$ formulation is given by

$$D^k_{\mathrm{DES}} = \rho \frac{k^{3/2}}{\bar{l}}, \tag{6.111}$$

where the length scale

$$\bar{l} = \min(l_{k-\omega}, C_{\mathrm{DES}}\Delta), l_{k-\omega} = \frac{k^{(3/2)}}{(\beta^*\omega)}, \tag{6.112}$$

where β^* has the same unit as k.

In order to find the optimal value of C_{DES}, the models were exercised in pure LES mode on a homogeneous decaying turbulence model, with the differencing scheme centered and fourth-order accurate, and the time integration was by an implicit second-order-accurate scheme. In practice, values of $C_{\mathrm{DES}} = 0.65$ for the single turbulence model, 0.67 for the $k - \varepsilon$ turbulence model, and 0.78 for the $k - \omega$ turbulence model was taken.

Around the year 2000, Travin et al. [117] also published their paper on DES simulation past a circular cylinder for flows with *laminar separation* (LS) at *Reynolds numbers* of 50,000 and 140,000, and with *turbulent separation* (TS) at 140,000 and 3×10^6, the latter separation having been done by effectively tripping the flow and also compared with the untripped flow at high *Reynolds numbers*. The finest grid had about 18,000 points in each of the grid planes spanwise; the resolution is far removed from DNS, and the turbulence model controlled the separation, if it was turbulent. The agreement was quite good for *drag, shedding frequency, pressure*, and *skin friction*. "However the comparison was obscured by large modulations of the vortex shedding and drag which were very similar to those seen in experiments but also, curiously, durably different between cases especially of the LS type. The longest simulations reached only about 50 shedding cycles. Disagreement with experimental *Reynolds stresses* reached about 30%, and the length of the *recirculation bubble* was about double that measured [117]." The technique follows that given by Spalart [105, 106] and Strelets [109].

It needs to be mentioned that a new DES method that is becoming popular, called *implicit DES method* (IDES), is designed to allow people to do unsteady simulations of boundary layer flows (Shur et al. [100]). Here, as given in the abstract of the paper, it "is proposed that combines *delayed detached-eddy simulation* (DDES) with an improved RANS-LES hybrid model aimed at wall modeling in LES (WMLES).

The system ensures a different depending on whether the simulation does or does not have inflow turbulent content. In the first case, it reduces to WMLES: most of the turbulence is resolved except near the wall. Empirical improvements to this model relative to the pure DES equations provide a great increase of the resolved turbulence activity near the wall and adjust the resolved logarithmic layer to the modeled one, resolving the issue of *log-layer mismatch*, which is common in DES and other WMLES methods. An essential new element here is a definition of the *subgrid length scale*, which not only depends on the grid spacings but also on the wall distance. In cases without inflow turbulent content, the proposed model performs as DDES, that is, it gives a pure RANS solution for attached flows and a DES-like solution for massively separated flows. The coordination of the two branches is carried out by a blending function. The promise of the model is supported by its satisfactory performance in all three modes it was designed for, namely, in pure WMLES applications (channel flow in a wide Reynolds number range and flow over a hydrofoil with trailing-edge separation), in a natural DDES application (an airfoil in deep stall), and in a flow where both branches of the model are active in different flow (a backward-facing-step flow)." For details please see the paper itself.

6.4.5 Large Eddy Simulation

The perfectly suited method to compute the large-scale fluctuations that are known to contribute to the noise generated in many problems is the LES technique. LES, with lower computational cost, is an attractive alternative to DNS, because in an LES, large scales are directly resolved and the effects of the small scales or the SGSs are modeled. The large scales are generally much more energetic than the small ones and are directly affected by the boundary conditions. The small scales, however, are usually much weaker and tend to have more or less a universal character. For high Reynolds numbers, the method allows prediction of the dynamics of the large turbulent scales, whereas the effect of the fine scale is modeled using a SGS model. The underlying assumptions for LES are based on a scale separation, the smallest scales of the exact solution being parameterized via the use of a statistical model referred to as a *Kolmogorov SGS model*.

"The most general methodology for the prediction of the far field noise is to compute the near field unsteady flow field using DNS or LES technique in conjunction with an acoustic analogy. Following the flow computation, an acoustic source term is then determined from the large scale fluctuating flow field. The far field acoustics are then calculated in a larger domain in grids, which are generally characterized by their coarser spatial resolution. As determined by the aeroacoustic problem considered different methods can be applied" [12].

One of the latest books on LES for incompressible flows is the third edition of the book by Sagaut [9].

Among the various applications of LES to aeroacoustics, some examples (not exhaustive) are given in what follows.

First was the problem of decaying and forced isotropic turbulence [101, 102], who addressed the modeling of the contribution of unresolved scales to radiated noise production when *Lighthill's analogy* was employed together with LES. To achieve this, they split *Lighthill's tensor* into three parts: a high-frequency part that is not resolved to LES, a filtered Lighthill tensor computed from filtered LES variables, and, finally, evaluating the subgrid and high-frequency contributions to complete the Lighthill tensor. This work showed that the high-frequency part of the Lighthill tensor does not contribute significantly to the noise production if cutoff numbers of the usual values are employed, but the work also indicated that the SGS contribution cannot be neglected. These authors investigated further whether the noise radiation generated in forced isotropic turbulence could be estimated using a hybrid LES–Lighthill-analogy approach and confirmed that the high-frequency part of Lighthill did not contribute significantly to the overall noise production and that the SGS intensity and the SGS pressure could be neglected. Also, they addressed the *parameterization of SGS effects* based on a *scale-similarity model* and demonstrated the efficiency of the latter in a priori and posterior tests [12].

The prediction of supersonic jet noise is an area that largely benefits from the recent developments [70, 71, 76], where a parallel three-dimensional computational aero-acoustic LES method is used with non-linear disturbance performed [82] using the SGS *eddy viscosity model*. Kolbe et al. [54] and Chyczewsky et al. [31] carried out LES of a rectangular jet but they relied on numerical and artificial damping to dissipate the turbulent energy. Mitchel et al. [77] performed direct numerical simulations for both the near-field flow and the far-field radiated sound from subsonic and supersonic 2D axisymmetric jets. Work was also done on jet flow and noise generation by *Boersma and Lele* [24] by conducting 3D LES of a sound jet under the same conditions as Freund [37] by using a dynamic Smagorinsky model [103], but no sound radiation predictions were reported. Additionally, *Mankbadi et al.* [72] studied supersonic jet noise using 2D LES witha *Smagorinsky model*, but the computation part did not include part of the acoustic field. *Bogey et al.* [26] conducted LES of a jet for a large *Reynolds number* and *Mach number* 0.9 with the standard *Smagorinsky model* and computed aerodynamic noise data like mean flow and turbulence intensities, as well as the *sound radiation directivity* and *sound levels* were in good agreement with the experimental data. Wang and Moin [126] performed LES computations of the flow past an asymmetrically beveled trailing edge of a flat strut at fairly large *Reynolds numbers*, and their computed mean and fluctuating velocity profiles compared reasonably well with experimental results. The unsteady incompressible flow around a blunt trailing edge has also been the subject of LES of turbulent flow around NACA 0012 airfoil for a Reynolds number RE = 1,000 based on the free-stream velocity and plate and plate thickness. Troff et al. [118] studied the problem by solving the *filtered Navier-Stokes equations* using an eddy viscosity SGS model on a non-staggered grid with a second order accurate hybrid finite difference-finite volume model.

"Large eddy simulation, with lower computational cost, is an attractive alternative to DNS. In an LES, the large scales are directly solved and the effect of the small scales or the SGSs or the SGSs on the large scales are modeled. The large scales are generally much more energetic than the small ones and are directly affected by the boundary conditions. The small scales, however, are usually much weaker and they tend to have more or less a universal character. Hence, it makes sense to directly simulate the more energetic large scales and model the effect of the small scales" [121].

Uzun et al. [119] have summarized recent progress in the creation of LES code for jet acoustics with the help of recently developed LES code as part of a CAA methodology that would eventually couple near-field unsteady LES data with an integral acoustic formulation for the prediction of far-field noise of turbulent jets. The code employs high-order accurate compact differencing together with implicit spatial filtering and state-of-the-art *nonreflecting boundary conditions*. The classical Subgirinsky subgrid-scale (SGS) model was used for representing the effect of unresolved scales on resolved scales. Preliminary results obtained in the formulations were encouraging when compared with the experimental data.

In LES, the governing equations are filtered for a certain length scale, and this results in a residual turbulent stress. On the other hand, one could perform such simulations without any residual turbulence model by invoking a dissipative mechanism that sets the smallest scale [69].

For this purpose we start by writing down first the spatially Favre-filtered continuity, momentum, and energy equations for compressible flow equations [15]:

$$\frac{\partial \overline{\rho}}{\partial t} + \frac{\partial (\overline{\rho}\tilde{u}_i)}{\partial x_i} = 0, \tag{6.113}$$

$$\frac{\partial \overline{\rho}\tilde{u}_i}{\partial t} + \frac{\partial (\overline{\rho}\tilde{u}_i\tilde{u}_j)}{\partial x_i} = -\frac{\partial \overline{p}}{\partial x_i} + \frac{\partial \overline{\sigma}_{ij}}{\partial x_j} + \frac{\partial \tau_{ij}}{\partial x_j}, \tag{6.114}$$

$$\frac{\overline{\rho}\tilde{e}_o}{\partial t} + \frac{\partial (\overline{\rho}\tilde{e}_o)}{\partial (\tilde{u}_j)}\frac{\partial (\overline{\rho}\tilde{u}_j)}{\partial x_j} = -\frac{\overline{p}\tilde{u}_j}{\partial x_j} + C_p\frac{\partial}{\partial x_j}\frac{\mu}{Pr} + \frac{\mu_t}{Pr_t}\frac{\partial \overline{T}}{\partial x_j} + \frac{\partial}{\partial x_j}\overline{u}_i(\overline{\sigma}_{ij} + \tau_{ij}). \tag{6.115}$$

Here $\overline{\sigma}_{ij}$ and τ_{ij} are the *Favre-filtered viscous stress tensor* and *SGS viscous stress tensor*, respectively. These are given by

$$\overline{\sigma}_{ij} = \mu\left(2\overline{S}_{ij} - \frac{2}{3}\overline{S}_{mn}\delta_{ij}\right) \tag{6.116}$$

$$\tau_{ij} = \mu_t\left(2\overline{S}_{ij} - \frac{2}{3}\overline{S}_{mn}\delta_{ij}\right) - \frac{2}{3}\overline{\rho}k^{SGS}\delta_{ij}, \tag{6.117}$$

where k^{SGS} is the *SGS kinetic energy*, μ_t is the *SGS kinematic viscosity*, and $\overline{S}_{ij}$ is the *Favre-filtered strain rate tensor* defined as follows:

$$k^{SGS} = C_I \Delta^2 \overline{S}_{mn} \overline{S}_{mn}, C_I = 0.66$$

$$\mu_t = C_R \overline{\rho} \Delta^2 (\overline{S}_{mn} \overline{S}_{mn})^{1/2}, C_R = 0.12 \text{ and}$$

$$\overline{S}_{ij} = \frac{1}{2}\left(\frac{\partial \overline{u}_i}{\partial x_j} + \frac{\partial \overline{u}_j}{\partial x_i}\right).$$

For SGS kinematic viscosity, a slightly different model [26] is

$$\mu_t = \overline{\rho}(C_s \Delta)^2 \sqrt{2 S_{ij} S_{ij}}, \tag{6.118}$$

where $C_s = $ the *Smagorinsky constant* and is taken to be equal to 0.18. Subsequent processing of the numerical results is done by computing space–time correlations for axial velocity with spatial separation ξ in the axial direction and a fixed time separation τ from the relation

$$R_{uu}(\mathbf{x}, \xi, \tau) = \frac{\overline{u'(\mathbf{x},t)u'(\mathbf{x}+\xi,t+\tau)}}{\sqrt{\overline{u'^2(\mathbf{x})}}\sqrt{\overline{u'^2(\mathbf{x}+\xi)}}}, \tag{6.119}$$

and further by computing the length and time scales of the turbulent properties. The time history of the pressure fluctuation in the observer location is obtained numerically from the LES in combination with Kirchhoff surface integration, from which the sound pressure level is obtained.

Finally, it is worth mentioning that the *implicit LES (ILES) method* has become very common these days for things like jets.

6.4.6　*Reynolds-Averaged Navier–Stokes Equations*

For turbulent flow predictions based on the *Reynolds-averaged Navier–Stokes equations* (RANS) additional closure equations are required, such as the standard $k - \varepsilon$ model (6.46) and (6.47), where k is the *turbulent kinetic energy*, and ε is the *turbulent dissipation*. Typically, k and ε are expressed in terms of the *turbulence intensity* and the *length scales* of the energy-carrying vortices.

The method relies on a statistical average of a "straightforward, steady RANS computation that provides information about turbulent length and time scales that translate by empirical relations into sound source spectra [12]. The relevant theorem says that the procedure can be asymptotically interpreted as a time-averaging procedure, leading to steady computations in the general case. Unsteady RANS can also be obtained when the statistical average is related to a conditional or phase-averaging procedure, or both. Note that the RANS approach does not permit explicit control of the complexity of the simulation because the cutoff frequency cannot be specified during the averaging procedure.

Some recent proposals for generating a field of synthetic or *artificial velocity fluctuations* from a given set of turbulence typical of those obtained from a classical RANS method have been made.

6.4.7 RANS-LES Simulation

Quemere and Sagaut [92] proposed coupling between two RANS and LES approaches. The authors consider a global RANS coupling exit as follows:

"The information provided by the RANS solver is used to derive some boundary conditions for the LES domain. To take into account the discontinuity between the two fields, the authors have proposed a two-way procedure to synthesize turbulent fluctuations. The first one relies on an extrapolation of the fluctuating variable computed in the LES region in the ghost cells of the LES variable given by RANS. The authors indicate that such a treatment is only valuable in the case of *lateral* or *outflow boundaries*, while for the *inflow boundaries* the authors have proposed to use a predictor simulation.

"A feedback from the LES region to the *steady RANS* or *unsteady URANS* region also exists. This is achieved by averaging the LES field to provide data for the RANS field boundary condition in the overlap region, where the traditional RANS transport equations are used to determine k, ε and other variables [12]."

Combining the method for synthesizing the turbulent RANS data that produce the second moments as well as the two-point correlations with the LES data, applied to RANS-LES interfaces, Batten et al. [21] performed a RANS-LES simulation, where "progress toward a general purpose hybrid Reynolds-averaged Navier–Stokes (RANS) with Large Eddy Simulation (LES) framework is described, in which large-scale, statistically represented turbulence kinetic energy is converted automatically into resolved scale velocity fluctuations wherever the local mesh resolution is sufficient to support them. Existing hybrid RANS/LES approaches alter the nature of the local partial differential equations according to the local mesh resolution, but they do not alter the nature of the data on which these equations operate. Implications of these are discussed. Subsequently, a simple mechanism is introduced to transfer statistical kinetic energy into resolved scale fluctuations in a manner that preserves a given set of space/time correlations and set of second moments. This process generates the large-scale eddies needed to form the unsteady boundary layer conditions at RANS interfaces to LES regions, into which the turbulence energy can be deposited either through mean convection or through turbulent transport. The proposed approach is designed to work on general meshes with arbitrary clusterings and does not require the user to specify internal boundary surfaces surfacing RANS and LES regions."

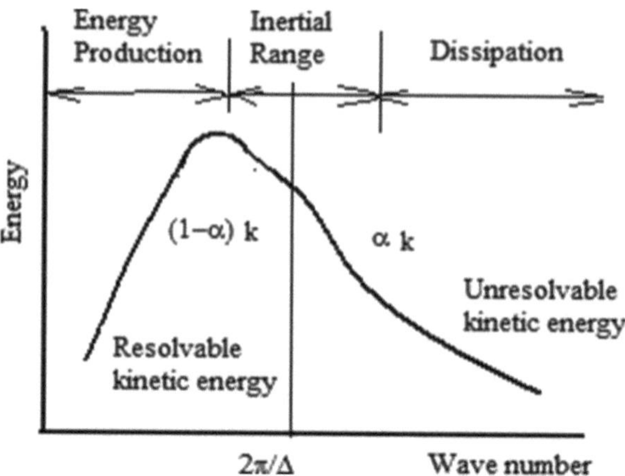

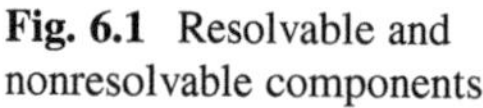

Fig. 6.1 Resolvable and nonresolvable components

The hybrid network proposed by Batten et al. [21] is the limited numerical scales (LNS) approach and was developed as a first application of the RANS/LES approach in a commercially available code.

One of the earliest mentions of a *hybrid RANS-LES method* was in a 1975 paper by Schumann [98], applied to the lower section of a boundary layer, however, but more recent methods typically envision treating the entire boundary layer with a RANS. The value of α is determined as the minimum of that in the LNS approach and RANS approach, as follows:

$$\alpha = \frac{\min[(L,V)_{\text{LES}},(L,V)_{\text{RANS}}]}{(L,V)_{\text{RANS}}}. \tag{6.120}$$

Using the definition of α in conjunction with 6.124, the governing equations behave as RANS if $\alpha = 1$, or as LES if $\alpha < 1$. When fine grid regions are encountered by the LNS method, the scale of the predicted *Reynolds stress tensor* by α causes the effective viscosity to be instantly reduced to the levels implied by the underlying *LES subgrid model*, with the local flow turbulence also experiencing a decreased rate of *stress-tensor components*.

The energy fraction αk is interpreted as unresolvable subgrid turbulent kinetic energy, which can only ever be modeled. The $(1 - \alpha)k$ is interpreted as resolvable turbulent kinetic energy, which could be represented directly on the local mesh (Fig. 6.1). In LNS, the sum total of statistically represented turbulence energy k_{LNS} does not have the same meaning as k_{RANS} in a traditional RANS closure. In general, $k_{\text{LNS}} \leq k_{\text{RANS}}$.

Similarly, the quantity $\alpha \varepsilon$ is interpreted as the dissipation applied to the nonresolvable scales, and the quantity $(1 - \alpha)\varepsilon$ is interpreted as the dissipation or transfer rate, which applies to the resolvable scales.

As mentioned earlier, a reasonable scale is one where the wavelength $\leq 2\Delta$, where Δ is the mesh spacing and 2Δ is the Nyquist limit of the mesh. Any wavelength shorter than this cannot be resolved directly and is modeled using some form of the sub-grid treatment.

In this work, the LNS model equations for the unresolved stresses are based on a slightly modified nonlinear $k - \varepsilon$ model, in which the *Reynolds stress tensor* is defined via a tensoral expansion, cubic in the mean strain and vorticity tensors, as follows: model, until the more recent papers of Speziale [107] and Spalart et al. [105] and others:

$$\frac{\partial \bar{\rho} \bar{k}}{\partial t} + \frac{\bar{\rho} \bar{u} \bar{k}}{\partial x_i} = \frac{\partial}{\partial u_i} \left[\left(\mu + \frac{\mu_t}{\sigma_k} \right) \frac{\partial \bar{k}}{\partial x_i} \right] + P_k - \bar{\rho} \bar{\varepsilon}, \tag{6.121}$$

$$\frac{\partial \bar{\rho} \bar{\varepsilon}}{\partial t} + \frac{\bar{\rho} \bar{u} \bar{\varepsilon}}{\partial x_i} = \frac{\partial}{\partial u_i} \left[\left(\mu + \frac{\mu_t}{\sigma_\varepsilon} \right) \frac{\partial \bar{\partial}}{\partial x_i} \right] + \left(C_{\varepsilon 1} P_k - C_{\varepsilon 2} \bar{\rho} \bar{\varepsilon} + E \right) T_t^{-1}, \tag{6.122}$$

where

$$P_k = -\bar{\rho} \overline{u_i'' u_j''} \frac{\partial \bar{u}_i}{\partial x_j},$$

$$\bar{\rho} \overline{u_i'' u_j''} = \alpha \bar{\rho} \frac{2}{3} \bar{k} \delta_{ij} - \mu_t S_{ij}^* + c_1 \frac{\mu_t \bar{k}}{\bar{\varepsilon}} \left(S_{ik}^* S_{kj}^* - \frac{1}{3} S_{kl}^* S_{kl}^* \delta_{ij} \right)$$

$$+ c_2 \frac{\mu_t \bar{k}}{\bar{\varepsilon}} \left(\Omega_{ik} S_{kj}^* + \Omega_{jk} S_{ki}^* \right) + c_3 \frac{\mu_t \bar{k}}{\bar{\varepsilon}} \left(\Omega_{ik} \Omega_{jk} - \frac{1}{3} \Omega_{lk} \Omega_{lk} \delta_{ij} \right)$$

$$+ c_4 \frac{\mu_t \bar{k}^2}{\bar{\varepsilon}^2} \left(S_{ki}^* \Omega_{lj} + S_{kj}^* \Omega_{li} \right) S_{kl}^*$$

$$+ c_5 \frac{\mu_t \bar{k}^2}{\bar{\varepsilon}^2} \left(S_{mj}^* \Omega_{il} \Omega_{lm} + S_{il}^* \Omega_{lm} \Omega_{mj} - \frac{2}{3} S_{lm}^* \Omega_m \Omega_{nl} \delta_{ij} \right)$$

$$+ c_6 \frac{\mu_t \bar{k}^2}{\bar{\varepsilon}^2} S_{ij}^* S_{kl}^* S_{kl}^* + c_7, \tag{6.123}$$

with

$$S_{ij}^* = \left(\frac{\partial \bar{u}_i}{\partial x_j} + \frac{\partial \bar{u}_j}{\partial x_i} \right) - \frac{2}{3} \frac{\partial \bar{u}_k}{\partial x_k} \delta_{ij},$$

$$\Omega_{ij} = \left(\frac{\partial \bar{u}_i}{\partial x_j} - \frac{\partial \bar{u}_j}{\partial x_i} \right),$$

$$S = \frac{\bar{k}}{\bar{\varepsilon}} \sqrt{\frac{1}{2} S_{ij}^* S_{ij}^*}, \quad \Omega = \frac{\bar{k}}{\bar{\varepsilon}} \sqrt{\frac{1}{2} \Omega_{ij}^* \Omega_{ij}^*}.$$

"Global (or nonzonal) hybrid RANS-LES methods rely on a single set of model equations and a continuous treatment that blends between RANS and LES approaches" [12]. *Speziale* presented a hybrid framework in which the stress tensor $\overline{u_i' u_j'}^M$ provided by a conventional RANS model was damped via

$$\overline{u_i' u_j'} = \alpha \overline{u_i' u_j'}^M, \tag{6.124}$$

where α is the LNS *latency parameter*

$$\alpha = \left[\exp^{-\beta L^\Delta/L^k}\right]^n,\tag{6.125}$$

where β and n are some *unspecified parameters*, L^Δ is some representative mesh spacing, and L^k is the *Kolmogorov length scale*.

The superscript "*M*" denotes values computed using commercial RANS applications from which the regular RANS stress levels are recovered whenever L^Δ is much larger than L^k, whereas the subgrid stresses vanish completely as $L^\Delta \to 0$.

Where β and n are some parameters, which were unfortunately never completely specified by Speziale [107] that would satisfy the two RANS and DNS limits, and all such choices guarantee that a suitable SGS (LES) model will be recovered in between. In addition, L^Δ is some representative mesh spacing and L^k is the *Kolmogorov length scale*.

A *realizable time scale* is defined as

$$T_t = (\bar{k}/\bar{\varepsilon})\max(1,\xi^{-1}),\tag{6.126}$$

where $\xi = \sqrt{(R_t)}/C_\tau$ with $R_\tau = \bar{k}^2/(v\bar{\varepsilon})$ and $C_\tau = \sqrt{2}$. The eddy viscosity is defined as

$$\mu_t = \alpha C_\mu f_\mu \frac{\bar{\rho}\bar{k}^2}{\bar{\varepsilon}},\tag{6.127}$$

with

$$C_\mu = \frac{2/3}{A_1 + S + 0.9\Omega},$$

$$c_1 = \frac{3/4}{(1,000 + S^3)C_\mu}, \quad c_1 = \frac{15/4}{(1,000 + S^3)C_\mu},$$

$$c_3 = \frac{-19/4}{(1,000 + S^3)C_\mu}, \quad c_4 = -10C_\mu^2,$$

$$c_5 = 0, \ c_6 = -2C_\mu^2, \ c_7 = -c_6,$$

$$f_\mu = \frac{1 - \exp^{-A_\mu R_t}}{1 - \exp^{-\sqrt{R_t}}}\max\left(1, \frac{1}{\xi}\right),$$

$$E = A_{E\tau}\bar{\rho}_{\max}\left[\bar{k}^{1/2}, (v\bar{\varepsilon})^{1/4}\right]\sqrt{\bar{\varepsilon}T_t}\Psi_\tau,$$

$$\Psi_\tau = \max\left(\frac{\partial \bar{k}}{\partial x_j}\frac{\partial \tau}{\partial x_j}, 0\right), \ \tau = \frac{\bar{k}}{\bar{\varepsilon}}.\tag{6.128}$$

Using the definition of α in (6.120), a *latency factor* α is defined for the current choice of RANS and LES component models as

$$\alpha = \min\left[\frac{C_s(L_i^\Delta)^2\sqrt{S_{kl}^*S_{kl}^*/2}}{C_\mu\bar{k}^2/\bar{\varepsilon}^2 + \delta}, 1\right],\tag{6.129}$$

with δ as a small parameter, $O(10^{-20})$, to allow $\alpha \to 1$ without singularities at low Reynolds number regions. The remaining model constants are

$$A_1 = 1.25 \, , \, C_{\varepsilon 1} = 1.44, C_{\varepsilon 2} = 1.92, \sigma_k = 1.0,$$

$$\sigma_\varepsilon = 1.3 \, , \, A_\mu = 0.01, A_{E1} = 0.15, C_S = 0.05. \tag{6.130}$$

The important distinguishing feature of LNS is that it contains no empirical constant beyond those appearing in the baseline RANS and LES models.

6.4.8   Aeroelastic-Acoustic Simulation

Turbulence not only causes noise, but it can also cause *aeroelastic problems* resulting in *creep and fatigue failures* in aircraft. It is a surface phenomenon. Also the turbulence in fluids is a volumetric phenomenon, and aeroelasticity is a surface phenomenon.

Gupta et al. [45, 46] have discussed "the details of a novel numerical *finite element based method* and a resulting code for the simulation of the acoustics phenomenon arising from aeroelastic interactions. Both computational fluid dynamics and structural simulations are based on finite element discretization employing unstructured grids. The sound pressure level on structural surfaces is calculated from the root mean square of the unsteady pressure, and the acoustic wave frequencies are computed from a *fast Fourier transform* of the unsteady pressure distribution as a function of time. The newly developed tool proves to be unique, as it is designed to analyze complex practical problems involving computations in a routine fashion."

The analysis starts with a finite-element FE structural modeling, followed by a free vibration analysis that computes the *natural frequencies* ω and modes φ by solving the matrix equation

$$M\ddot{u} + Ku = 0, \tag{6.131}$$

where M and K are the inertia and stiffness matrices of order N and u is the *displacement vector*.

A steady-state fluid flow solution is next derived using a *two-step solution procedure* and local time stepping; this is accomplished by using the *time-dependent Navier–Stokes equation*

$$\frac{\partial v}{\partial t} + \frac{\partial f_j}{\partial x_j} + \frac{\partial g_j}{\partial x_j} = f_b \, , \, j = 1, 2, 3. \tag{6.132}$$

Once the aerodynamic parameters are calculated, the vehicle equation of motion is then cast in the frequency domain as shown by

$$\hat{M}\ddot{q} + \hat{C}\dot{q} + \hat{K}q + f_a(t) + f_l(t) = 0, \tag{6.133}$$

where q is the generalized displacement vector and $\hat{M}$ is the *generalized inertial matrix*, a similar transformation is adopted for the stiffness $\hat{K}$ and damping $\hat{C}$ matrices, f_a is the *aerodynamic load vector*, and $f_l(t)$ is the *impulse force vector*.

From the general analysis with acoustic simulation, the unsteady pressures are computed, followed by the average pressure and the root mean square of the pressure fluctuations, which gives the *sound pressure level*. Similarly the acoustic-wave frequencies are derived performing *fast Fourier transform* on the unsteady pressure data.

In finite-difference computation of compressible flow, it is necessary to have grids aligned to the flow, and in many cases the upward difference schemes are among the stable cases. Similar problems in the FE method were overcome by coordinate transformation of the surface elements to make sure the transformed elements are aligned with the flow direction.

In order to verify the solution accuracy, studies were done on a simple NACA 0012 airfoli (symmetric airfoil with zero camber and 12% maximum thickness-to-chord ratio), and subsequently a 3D cantilever wing with a NACA 0012 airfoil was analyzed in a O-type grid. The analysis starts with a steady-state CFD solution, followed by a structural modal analysis. Also NASA Sofia aircraft data for aircraft flying at 10 km altitude and 323 knots were used for comparison. The resulting aerodynamic and structural dynamic data are then used for the aeroelastic formulation.

6.5   Propeller Noise

Propellers have existed for a long time as a driving force on ships, aircraft, and windmills. The first aircraft designed by the Wright brothers in 1904 in their cycle shop had as a driving engine a small multiple-cylinder single-lined gasoline engine. Subsequently, increasingly larger multiple-cylinder radial engines to drive propellers were designed for bigger aircraft, but even now small aircraft have propellers.

In the 1940s, during the Second World War, jet engines were designed for military applications in England and Germany, and later, in the 1950s, the first commercial jet airliner called Comets were designed, although propeller-driven big commercial aircraft, such as Constellation and Super Constellation-G, flew for a few more years.

For propeller-driven engines, the air exhaust speed is just above the flight speed of the aircraft, while for jet engines the former is much larger than the flight speed. Therefore, as can be shown from the theory of aircraft engines (giving exact expressions would be quite out of place for the present discussion), the propeller-driven engines have a much better overall efficiency and as a consequence lower *specific fuel consumption* (fuel mass-flow rate per unit of thrust) than jet engines.

Michael Carley [29] of Trinity College in Dublin (Ireland) wrote his doctoral thesis in 1996 regarding the time-domain calculation of noise generated by a propeller in a flow. "The problem of calculating the noise from an acoustically subsonic aircraft propeller is approached using a *moving medium* method. An established theory of noise generation by rigid bodies in motion is combined with the moving medium *Green's function* to develop a linear acoustic formulation for the sound radiated by a body in arbitrary motion in a uniform steady flow of arbitrary orientation. Results from a numerical code based on this formulation are then presented and compared to experimental results from a test campaign conducted as part of the EU sponsored *Study of Noise and Aerodynamics of Advanced Propellers* project. Near field acoustic results from two propellers – a conventional *low speed* propeller and an advanced *high speed* design are presented for a range of operating speeds in the high subsonic to supersonic range. Further results are presented for the far field noise of the low speed propeller operating at a low flight Mach number. In each case the blade loading distribution used in the calculations is interpolated from experimental data. It is found that the numerical code predicts the experimental data quite well but is more accurate in predicting the noise from the high speed propeller than the low speed design, even for supersonic blade tip speeds (up to Mach number 1.8)." The results are computed and presented for various *chord angles* from $54.9°$ to $61.8°$, *advance ratio* from 3.238 to 4.625, *approaching flow Mach numbers* from 0.2 to 1.08, ad *tip Mach number* from 0.84 to 0.93, and at different radius to hub radius. Predicted acoustic pressure from different radii generally agree well with experiments.

Further, Marte and Kurtz [73] of *Jet Propulsion Laboratory* have presented a review of aerodynamic noise from propellers, rotors, and lift fans, with respect to *rotational noise, vortex noise, turbulence-induced noise*, etc.; in addition, formulas for calculation of the sound pressure level (SPL) as a function of nondimensional parameters at a prescribed distance of these various types of blade noise are given by them. The expression indicated that the *vortex noise* is a strong function of blade velocity (at specified 0.7 radius) and propeller blade area, and a doubling of the blade velocity increases the SPL by 18 dB.

Along with this topic on propeller noise, let us address also the question of *wind turbine noise*. Alberts [14] of Lawrence Technological University has asked this very question. As an introduction, he mentions that Michigan, in its *Renewable Energy Program*, had a goal of installing 800 MW of wind power by 2010. "Wind turbines generate two types of noise: aerodynamic and mechanical. A turbine's sound power is the combined power of the both. Aerodynamic noise is generated by the blades passing through the air. The power of aerodynamic noise is related to the ratio of the blade tip speed to wind speed," as indicated in Table 6.1. Usually wind turbines supplying electric power to electricity grids must have constant r.p.m., and at different wind speeds the blade angles have to be changed. Therefore, for the values given in Table 6.1, it is assumed that the blade tip speed remains constant and only wind speed changes.

Depending on the turbine model and the wind speed, the aerodynamic noise may seem like buzzing, whooshing, pulsing and even sizzling. Turbines with their blades downwind of the tower are known to cause thumping sound as each blade passes

Fig. 6.2 Wind turbines in Californian desert

Table 6.1 Sound power of small wind turbines

Make and model	Turbine size	Wind speed (ms^{-1})	Estimated sound power (dB)
Whisper H400	900 W	5	81.8
		10	91
Bergy Excel BW03	10 kW	5	87
		7	96.1
		10	105.4

Table 6.2 Sound power of some utility turbines

Make and model	Turbine size	Sound power
Vestes V80	1.8 MW	98–109 dB
Enercon E70	2.0 MW	102 dB
Enercon E112	4.5 MW	107 dB

the tower. Most noise radiates perpendicular to the blade rotation. However, since the blades rotate to face the wind, they may radiate noise at different directions each day. The noise from two or more turbines may combine to create an oscillating or thumping "wa-wa" effect (Fig. 6.2).

"Wind turbines generate broadband noise containing frequency components from 20 to 3,600 Hz. The frequency composition varies with wind speed, blade pitch, and blade speed. Some turbines produce noise with a high percentage of low-frequency components at low wind speeds than at high wind speeds [14]".

Utility-scale turbines must generate electricity that is compatible with grid transmission (not if the wind power is for pumping water or for grinding mills, as in the Netherlands). To meet this requirement, turbines are programmed to keep the blades rotating at as constant a speed as possible. To compensate for minor wind speed changes, the pitch of the blades is adjusted, change the sound power levels and frequency components of the noise. Table 6.2 lists the sound power for some common utility-scale turbines.

In this section, we would also like to mention the case of a *Dyson fan*.

These fans have a cylindrical base with a fan and holes through which air is sucked into the cylinder and pushed into a ring airfoil using a vortex effect to accelerate it to 15 times its original speed; it exits near the leading edge of the ring airfoil. This air flow is sucked into the ring and, by ejector action, brings in about 15 times the mass flow of the original flow in a very smooth manner without buffeting. One can put a hand inside the ring without breaking anything. Thus, this is a perfect solution for a child's room.

Dyson has produced a very beautiful object with a successful – nonmetallic finish – in two colors: white or gray. It can be touched and tilted easily, and you can put your hand on the ring.

When the Dyson fan is switched on, its throughput can be adjusted via a central knob without any steps, it can be set to rotate automatically, and it can be tilted; however, the fan costs several times more than a normal bladed fan.

The fans are noisier than a standard fan. While standard fans may give off 50 dB of noise at the highest setting, a Dyson fan, under the same conditions, emits 59 dB of noise. However, at a lower setting the Dyson fans are very quiet and give off a nice breeze.

The Dyson fan consumes much less energy than other fans. At the lowest setting, the Dyson fan produces 51 dB of noise, the same as a normal fan, but consumes only 9 W, much less than other fans. For maximum rating, the noise level is 54 dB, but its energy consumption is a quite acceptable 14 W.

6.6 Helicopter Noise

"There are a variety of noise sources associated with helicopters, for example, helicopter external noise due to helicopter main and tail rotors, which includes steady, periodic and random loads on the rotor blades, as well as volume displacement and nonlinear aerodynamic effects at high blade Mach numbers. Either main or tail rotors can be dominant noise sources at various frequencies and observer positions. Engines also produce noise of various types. Turbines produce inlet, compressor, turbine, combustor and jet noise, and reciprocating engines produce intake and exhaust noise as well as noise due to structural vibrations. Gearbox vibration can also contribute to external and interior noise" [43, 57, 80].

For aerodynamic noise in helicopters, there are basically two types. The first is called *high-speed impulsive (HSI) noise* and is characterized by high tip speeds and advancing tip Mach numbers larger than 0.9. The second type comes from the interaction of the rotor blades with their vortical wake systems, which are specially important for helicopters descending for landings [108].

At large forward flight speeds, *shock-associated rotor noise* is a major noise source. However, presently there is no good method that gives full physical insight into the process, and it is most efficiently predicted using integral methods that separate the computation of the noise sources from the noise propagation. The aerodynamic field around the blade is evaluated using a suitable unsteady CFD solver, and an integral formulation is used to describe how the sound propagates to the far field. The two most commonly used integral methods are *Kirchhoff's method* and the *Ffowcs–Williams equation*. Kirchhoff's method involves integration over a surface located in the linear flow region without requiring any volume integration, but its drawback for transonic flows is that the linear flow region is typically far away from the blade [80].

The aerodynamic noise from main rotors can be classified into two main categories: rotational (harmonic) noise and broadband noise. The aerodynamic sources of rotational noise are mean lift and drag forces, harmonic force fluctuations, and wake self-noise. At large distances from the vehicle, the low-frequency noise from the main rotor dominates since the higher frequencies are absorbed by the atmosphere [57].

The AIAA meeting paper by Strawn et al. [108] discusses "several new methods to predict and analyze rotorcraft noise. These methods are: (1) a combined CFD and Kirchoff scheme for far-feld noise predictions, (2) parallel computer implementation of the Kirchoff integrations, (3) audio and video rendering of the computed acoustic predictions over large far-field regions, and (4) acoustic tracebacks to the Kirchoff surface to pinpoint the sources of the rotor noise." The paper describes each of these methods and presents results for three sample cases. "The first case consists of in-plane high speed impulse noise and the other two cases show idealized parallel and oblique blade-vortex interactions."

The Kirchoff formulation is used to evaluate the acoustic pressure, p, at a fixed observer location, $\mathbf{x}$, and observer time, t, on a *Kirchoff surface*, which can deform and have arbitrary motion, although in the paper a nonrotational, rigid surface with linear translation motion is used. With these simplifications the *Kirchoff formula* becomes

$$p(\mathbf{x},t) = \frac{1}{4\pi} \int_S \left[\frac{E_1}{r(1-M_r)} + \frac{E_2}{r^2(1-M_r)} \right] - r\mathrm{d}S, \qquad (6.134)$$

where

$$E_1 = (M_n^2 - 1)p_n + M_n\mathbf{M}_t \bullet \nabla_2 p - \frac{M_n \cdot \overset{\cdot}{p}}{a_\infty} + \frac{(\cos\theta - M_n)\overset{\cdot}{p}}{a_\infty(1-M_r)}, \qquad (6.135)$$

$$E_2 = \frac{(1-M^2)(\cos\theta - M_n)}{(1-M_r)^2}. \qquad (6.136)$$

The *Kirchoff surface* translates with the rotor hub at *Mach number* $\mathbf{M}$. The two helicopter rotor blades rotate inside the nonrotating Kirchoff surface. The distance between a point on the *Kirchoff surface* and an observer is given by $|r|$. M_n and M_r are the components of $\mathbf{M}$ along the local surface normal, $\mathbf{n}$, and the radiation direction, $\mathbf{r}$; $\mathbf{M}_t$ is the *Mach number* tangent to the *Kirchoff surface*; p_n is the derivative of p along the *surface normal*; $\overset{\cdot}{p}$ is the time derivative of the pressure; and $\nabla_2 p$ is the gradient of the pressure on the *Kirchoff surface*. The free-stream speed of sound is assumed at uniform a_∞, and θ is the angle between $\mathbf{n}$ and $\mathbf{r}$.

The integrand must be evaluated at the time of emission for each differential area of the Kirchoff surface, noting the *retarded time*, τ, at the observer. This formulation leads to a quadratic equation for τ, one of which gives the positive real solution, with the advantage that the solution of τ does not require any iteration. Results of the computation of high-impulsive noise when the pressure is plotted against the blade azimuthal angle shows very high agreement between the experimental and computed results.

The 2005 *AIAA J.* paper by Morgan et al. [80] discusses "Transonic Helicopter Noise," which is an increasingly important issue, and, at large forward-flight speeds, is a major contributor to "noise pollution. To perform the noise predictions, an Euler CFD method to calculate the flow field combined with an acoustic method incorporating retarded time formulation of the *Ffowcs–Williams (FW) equations* [39] was used, for which several rotor blades in hover and steady forward flight were considered".

The FW equation expresses noise in terms of a distribution of the monopole and dipole sources over the volume inside the surface. When the control surface is chosen to coincide with the blade surface, these distributions represent "the noise due to the blade thickness, the blade loading, and the flow nonlinearities/entropy variations, respectively. It has been shown that the noise generated by the volume quadrupole distribution is the subsonic volume of fluid, but it is significant in regions of transonic flow.

"To predict shock-associated noise while avoiding the need to compute the problematic quadrupole term, the FW equation can be applied to a permeable control surface that encloses the blade but is not coincident with it. If the control surface is also chosen such that it encloses all transonic regions of the flow, then the volume outside the control volume will be fully subsonic, and the noise generated by the quadrupole distribution outside of the surface will be negligible. Thus by moving the control surface outward, the effect of the quadrupoles can be accounted for by the surface source terms. Furthermore, the transonic region is always well defined and the surface source terms continue to have physical meaning; the monopole distribution is related to the mass flux through the surface and the dipole distribution to the momentum flux [80]."

The permeable form of the FW equation was applied for simple hover, nonlifting forward flight, and lifting forward flight at various blade tip Mach numbers up to 0.85, and the results of calculation were found to be in good agreement with experimental results.

6.7 Exercises

6.7.1 What are the differences between *computational fluid dynamics* (CFD) and *computational aeroacoustic* (CAA) solutions?

6.7.2 Enumerate the different noise-generating mechanisms with practical examples.

6.7.3 Write down the expression for the 3D *turbulent shear stress*.

6.7.4 What are *turbulent kinetic energy* and *turbulent energy dissipation*? Are these *scalar* or *vector* quantities?

6.7.5 What is *Lighthill's analogy*? State the principles on which it is based.

6.7.6 What is the *Kolmogorov scale* and how is it determined? What is its relevance in computing aerodynamic noise?

6.7.7 What are resolvable and nonresolvable problems? How are the latter solved?

6.7.8 What is a *Kirchoff surface*?

6.7.9 What is a *realizable time scale*?

Chapter 7
Further Topics in Aerodynamic Noise

7.1 Supersonic Jet Noise

In principle, the convected quadrupole theory, as developed by Ffowcs-Williams, can be used also for supersonic jet noise. However, subsonic jet noise, where no complex phenomena such as shock or *screeching noise* exist, is based mainly on turbulent mixing. Supersonic jet noise, on the other hand, "is a cumulative effect of Mach wave radiation, nozzle tip radiation, nozzle lip radiation, shock turbulence interaction, shock unsteadiness and turbulent mixing" [65]. Mach waves, generally for supersonic, fully expanded nozzles, are mainly due to disturbance-convected, supersonically radiating sound in a highly directional peak and originating in a region of a few jet diameters from the nozzle exit. For supersonic over- or underexpanded nozzles, noise is produced by turbulence shock interactions, and it occurs at highly discrete frequencies.

From experiments it is found that for a fully expanded supersonic jet, this can be divided into several regions (Fig. 7.1) in which the location of the peak noise is just within the supersonic region inside the sonic line. Experiments have shown that, only for parallel jets,

$$L_c/D_o = 5.22 M_{\text{jet}}^{0.9} + 0.22 \text{ for } M_{\text{jet}} > 0.7$$

$$= 4 \text{ for } M_{\text{jet}} < 0.7,$$

$$L_s/D_o = 5 M_{\text{jet}}^2 + 0.8.$$

For an *overexpanded supersonic jet*, there are cells of *oblique shocks*, giving rise to high-frequency *screeching noise* with characteristic frequency related to the shock cell lengths, which is different from the *broadband noise* of a correctly expanded or *underexpanded supersonic jet*.

For subsonic jets a fully developed region starts generally from $8D_o$. But, for a supersonic jet, the beginning of a fully developed turbulent jet with a similar profile will not occur until much farther downstream. Once the turbulent jet becomes fully developed, the available experimental data indicate that the velocity on the axis

T. Bose, *Aerodynamic Noise: An Introduction for Physicists and Engineers*,
Springer Aerospace Technology 7, DOI 10.1007/978-1-4614-5019-1_7,
© Springer Science+Business Media New York 2013

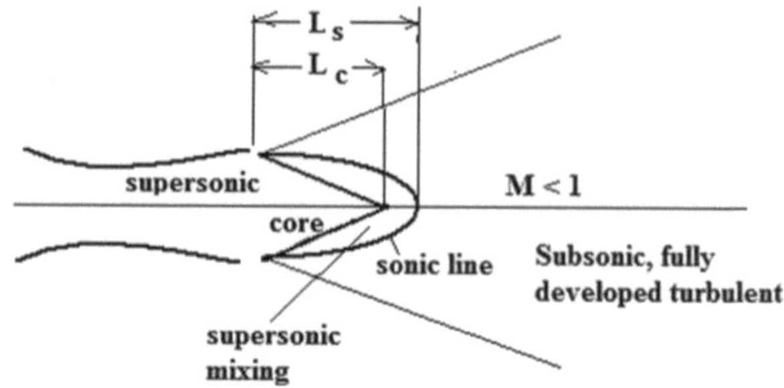

Fig. 7.1 Mixing regions for supersonic, well-expanded parallel jet

decays for a supersonic jet like x^{-1}. This rule seems to be valid even for originally supersonic jets because the fully developed subsonic jet is not established until the flow becomes subsonic.

Thus the basic differences between subsonic and supersonic jets are as follows:

1. The supersonic mixing region extends over a large distance, which for low subsonic jets is approximately four times the jet exit diameter.
2. The acoustic power output per unit length is not constant in the mixing region for a supersonic jet but increases by four times the jet exit diameter.
3. The acoustic power output in the subsonic region can be computed by the Lighthill theory.

Although Nagamatsu [83] proposed a semiempirical method for computing supersonic noise, he failed to explain the linear variation in the acoustic power per unit length in the supersonic region. *Tam* [110] proposed an independent idea according to which the dominant part of jet noise is generated not by random jet turbulence or eddy Mach waves, as is generally believed, but by orderly large-scale disturbances in the jet flow, which may be generated inside the settling chamber but may be propagated downstream through the nozzle. Bose [28] made a tentative proposal regarding this model as given in what follows.

The underlying assumptions for this analysis are as follows:

1. The flow is one-dimensional.
2. There are no viscous forces.
3. There is no axial heat flux by conduction.
4. The gas in the nozzle is ideal and frozen.
5. c_p and γ are constants.
6. Any fluctuations in the flow properties are isentropic, although the time-averaged properties may change in the flow direction with a polytropic exponent $n \geq \gamma$.

The two basic equations are

$$\text{Continuity: } (\rho A)_t + (\rho u A)_x = 0, \tag{7.1}$$

$$\text{Momentum: } (\rho u)_t + (\rho u^2)_x + p_x = 0. \tag{7.2}$$

Here x and t denote the partial derivatives with respect to x and t, respectively. In addition, a total temperature is defined locally by the relation

$$T_o = T + u^2/(2c_p). \tag{7.3}$$

Now all variables can be written in terms of the sum of time-independent variables and fluctuating quantities:

$$\rho = \bar{\rho} + \rho', u = \bar{u} + u',$$
$$p = \bar{p} + p', T = \bar{T} + T' \text{ and } T_o = \bar{T} + T_o'.$$

For constant stagnation temperature, $T_o = $ constant,

$$T_o = T + T' + (\bar{u}^2 + 2\bar{u}u' + u'^2)/(2c_p). \tag{7.4}$$

If separate equations are written for time-independent variables and time-dependent variables parts separately, and if the terms containing the square of the perturbed quantities are neglected, then we get the relation $T' = -\bar{u}u'/c_p$.

Further, from equations for isentropic change of state, for example $p^{(\gamma-1)/\gamma}/T = $ constant, we get by logarithming and diferentiating the following set of relations:

$$\frac{p'}{\bar{p}} = \frac{\gamma}{\gamma-1}\frac{T'}{T}; \frac{\rho}{\bar{\rho}} = \gamma\frac{p'}{\bar{p}},$$

$$T' = -\frac{\bar{u}u'}{c_p}; c^2 = \gamma\frac{\bar{p}}{\bar{\rho}},$$

$$u' = -\frac{c_p T}{\bar{u}} = -\frac{RT p'}{\bar{u}p} = -\frac{p'}{\bar{\rho}\bar{u}} = -\frac{Ap'}{\dot{m}_o}.$$

Thus,

$$\frac{\partial(\rho u)}{\partial t} = \frac{\partial}{\partial t}[(\bar{\rho} + \rho')(\bar{u} + u')] = \bar{\rho}\frac{\partial u'}{\partial t} + \bar{u}\frac{\partial \bar{\rho}}{\partial t}$$

$$= \left[\frac{\bar{u}}{c^2} - \frac{1}{\bar{u}}\right]\frac{\partial \rho'}{\partial t} = \frac{(\bar{M}^2 - 1)}{\bar{u}}\frac{\partial p'}{\partial t}.$$

Now (7.1) and (7.2) are differentiated with respect to x and t, respectively, and subtracted from one another. In addition, the variables are written as the sum of the time-averaged and perturbed variables, in which products of two fluctuating quantities are neglected, and finally we get the equation

$$\frac{\partial^2 p'}{\partial x^2} - \frac{1}{c^2}\frac{\partial^2 p'}{\partial t^2} = \frac{(\bar{M}^2 - 1)}{\bar{u}A}\frac{\partial A}{\partial x}\frac{\partial p'}{\partial t} = 0. \tag{7.5}$$

We define a function $\varphi(x)$ by the relation

$$\varphi = \frac{(\bar{M}^2 - 1)}{\bar{M}2\pi\nu}\frac{c}{A}\frac{\partial A}{\partial x},\tag{7.6}$$

where ν is the characteristic frequency of oscillation. It may be noted that in a convergent–divergent nozzle, in both the convergent (usually subsonic) and divergent (usually supersonic) sections, $\varphi(x)$ is always ≥ 0. Similarly, at the throat it is equal to zero, which is also the case for a constant tube cross section. In the latter case, (7.5) is reduced to a 1D wave equation with a solution

$$p' = P\exp^{j2\pi\nu|(x/c)-t|},\tag{7.7}$$

where P is the amplitude of the pressure fluctuation. On the other hand, the more general (7.5) is rewritten in the form

$$\frac{\partial^2 p'}{\partial x^2} - \frac{1}{c^2}\left(\frac{\partial^2 p'}{\partial t^2} + 2\pi\nu\varphi\frac{\partial p'}{\partial t}\right) = 0.\tag{7.8}$$

Now we seek a trial solution such that it will also satisfy the case of constant tube cross section. One such trial solution we try is

$$p' = P\exp^{j2\pi\nu}\left| \int c^{-1}\mathrm{d}x - \int(1+\lambda)\mathrm{d}t\right|,\tag{7.9}$$

which, when substituted into (7.8), gives the result

$$\lambda = -1 + 0.5\varphi + \sqrt{(1 - 0.25\varphi^2)}.\tag{7.10}$$

Thus λ is a complex variable, in which the real part gives a change in frequency and the imaginary part gives the amplification or the damping of the amplitude of pressure fluctuation known at a reference point $x = 0$. The real and imaginary parts of λ depend on the value of φ and are given by the following relations:

$$\varphi \leq 2 : \mathrm{Real}(\lambda) = -1 + \sqrt{1 - 0.25\varphi^2},\ \mathrm{Im}(\lambda) = 0.25\varphi,$$

$$\varphi \geq 2 : \mathrm{Real}(\lambda) = -1,\ \mathrm{Im}(\lambda) = 0.5\varphi \pm \sqrt{0.25\varphi^2 - 1}.$$

From the preceding set of results it can be seen that for all possible values of φ as positive, real (λ) is always negative and Im (λ) is always positive. Therefore, along a nozzle, except at the throat, the frequency is always reduced and the amplitude is also always damped. Further, whether φ is large or small depends on whether the frequency of fluctuation is small or large. From the calculated distribution of λ along the nozzle for a given frequency, while it is possible to obtain the extent to which the frequency and the amplitude of pressure fluctuations may vary along the nozzle, one can determine fluctuations of other flow variables.

At this point it can only be speculated whether the method is applicable also for understanding jet noise in the supersonic range.

7.2 Sound at Solid Boundaries

Physically, solid boundaries can make their presence felt in two ways:

1. The sound generated by quadrupoles are reflected and refracted by the solid boundaries.
2. The quadrupoles are no longer distributed over the entire space but only throughout the region external to the solid boundaries, and it seems that there may be a resultant distribution of dipole (or even sources) at the boundaries.

In the presence of solid boundaries, the most general solution of (4.5) with quadrupole terms alone (Stratton [11], Curle [35]) is

$$\rho - \rho_o = \frac{1}{4\pi c_o^2} \int_\Omega \frac{\partial^2 T_{ij}}{\partial x_i \partial x_j} \frac{d\Omega(\mathbf{y})}{|\mathbf{x} - \mathbf{y}|} + \frac{1}{4\pi} \int_S \left[\frac{1}{r} \frac{\partial \rho}{\partial n} + \frac{1}{r^2} \frac{\partial r}{\partial n} \rho + \frac{1}{c_o r} \frac{\partial r}{\partial \mathbf{r}} \frac{\partial \rho}{\partial t} \right] d\mathbf{S}(\mathbf{y}),$$

$$(7.11)$$

where $\mathbf{n}$ is the normal vector perpendicular to the surface element $d\mathbf{S}$. In the absence of solid boundaries, the surface integral will not appear and only the volume integral remains. This is not quite the same as (3.24). It is obtained, however, by considering the quadrupole field as four separate source fields as they come infinitely close together. While the effect of solid boundaries is given by the extra surface integral part in the preceding equation,

$$\frac{1}{\pi} \int_S \left[\frac{1}{r} \frac{\partial \rho}{\partial \mathbf{n}} + \frac{\rho}{r^2} \frac{\partial r}{\partial \mathbf{n}} + \frac{1}{c_o r} \frac{\partial \rho}{\partial \mathbf{n}} \frac{\partial \rho}{\partial t} \right] dS(\mathbf{y}). \qquad (7.12)$$

It must be noted, however, that the presence of fixed solid boundaries invalidates the idea of regarding the quadrupole distribution as the limiting case of a four-source distribution. Thus the physical idea is exactly equivalent to the mathematical process of twice applying the divergence theorem, and carrying out this process is still permissible. By complicated transformation of the integral Curle [35] gets the equation

$$\rho - \rho_o = \frac{1}{4\pi c_o^2} \frac{\partial^2}{\partial x_i \partial x_j} \int \frac{T_{ij}}{r} d\Omega(\mathbf{y}) - \frac{1}{4\pi c_o^2} \frac{\partial}{\partial x_i} \int \frac{F_i(\mathbf{y})}{r} d\mathbf{S}(\mathbf{y}), \qquad (7.13)$$

where T_{ij} and F_i are computed at the retarded time $(t - r/c_o)$. Now $F_i = -l_j \tau_{ij}^*$, where τ_{ij}^* is the stress tensor and l_j is the direction cosine of the outward normal from the fluid. Thus F_i is the maximum in the normal direction to the surface, and it is exactly the force per unit area exerted on the fluid by the solid boundaries in the x_i direction and has the same unit as τ_{ij}^* (Nm^{-2}). Physically, therefore, one can look upon the sound field as the sum of noise sources that are generated by a volume distribution of quadrupoles and by a surface distribution of dipoles.

Taking only the second term on the right-hand side of (7.13) for evaluation, one gets the following equations, which are derived in the same manner as previously in the case of quadrupole radiation, for the evaluation of the sound radiation in the far field

$$I(\mathbf{x}) = \frac{c_o^3 B(\mathbf{x}, \theta^* = 0)}{\rho_o}, \tag{7.14}$$

where

$$B(\mathbf{x}, \theta) = \overline{(\rho - \rho_o)^2}$$
$$= \frac{1}{16\pi^2 c_o^4} \int \int \frac{x_i x_j}{c_o^2 |\mathbf{x}|^4} F_i(\mathbf{y}, t) F_j(\mathbf{y} + \delta, t + \theta) \, d\mathbf{S}(\mathbf{y}) d\mathbf{S}^*(\delta). \tag{7.15}$$

In (7.15), $\mathbf{S}^*$ is the correlation surface, θ is the time delay at the source, and θ^* is the time delay at the observer. By assuming $F_i \sim \rho_o U^3 / x$ and $\mathbf{S} \sim x^2 U / c_o$, one gets the expression $I \sim \rho_o U^8 / c_o^5$. The radiated sound power per unit area of the surface for a rigid wall has been given by Ribner [8] at low speeds approximately by the relation

$$I = 5 \times 10^{-4} \rho_o U^8 / c_o^5, \ \text{Wm}^{-2}, \tag{7.16}$$

giving a U^8 law characteristic of quadrupole aerodynamic noise.

The unsteady boundary layer pressure, however, tries to force vibration of a flexible wall, where the pressure field is given empirically by the relation

$$I = 45.4 + 40 \log_{10} U + 20 \log_{10} \left(\frac{\rho}{\rho_{H=0}} \right), \ \text{dB} \tag{7.17}$$

where U is the subsonic characteristic velocity in microseconds. This rather high intensity is recorded by a microphone mounted flush in the wall under the boundary layer.

7.3   Combustion Noise

The explosive sound of piston engines due to periodic combustion is the major source of noise in a piston aircraft engine. The mechanism can be studied by assuming a monopole mechanism. If the volume flow is $\dot{\Omega}(t)$, then the density fluctuation is given by

$$\rho(t, r + r/c_o) - \rho_o = \rho_o \dot{\Omega}(t) / (4\pi c_o^2) \tag{7.18}$$

and the instantaneous *intensity of sound* and *acoustic power output* are given by the relations

$$I(r, t + r/c_o) = \frac{c_o^3 \overline{(\rho - \rho_o)^2}}{\rho_o} = \rho_o \frac{\overline{\dot{\Omega}^2(t)}}{16\pi^2 r^2 c_o}, \tag{7.19}$$

$$P(r, t + r/c_o) = \rho_o \frac{\overline{\dot{\Omega}^2(t)}}{4\pi c_o}. \tag{7.20}$$

For better accuracy, one can do the same kind of analysis as that of a pulsating piston in which the lower harmonics will be nondirectional, but at high frequencies there will be more radiation in the direction normal to the surface.

7.4 Sonic Boom

It is a common experience that for a supersonically flying aircraft sonic booms or bangs are heard on the ground. If the variation of pressure on the ground (*pressure signature*) is measured, it is a characteristically shaped *N-wave*. For production of these waves and for the simplest case of a homogeneous atmosphere, Whitham [128] modified the earlier linearized theory of supersonic flow and obtained results that reproduce the experimental results well.

According to *Whitham*, shocks that occur in regions where the characteristics would otherwise form a limit line and the solution would cease to be single valued are otherwise determined from simple geometrical property that, to a first order in its strength, a shock bisects the angle between the characteristics on each side of it. Thus, as a first approximation, isentropic equations of motion are used, not because it is thought that a good approximation will be made to the correct (nonisentropic) case since the shocks concerned are weak and the entropy changes at a shock are of the third order in its strength. It is found that the theory is correct to the second order with an important exception, that is, the position of the rear shock is correct only to a first order, which is due to a third-order pressure wave spread out behind the main disturbance over a large distance.

It is assumed in the theory that the body is slender and pointed at the nose with the first shock attached. For such cases and for a large distance from the body generating the shock in comparison to the characteristic cross-flow dimensions of a body, the deviation from a cylinder of constant radius may be found to be small. Hence the results for the front shock and all the theory at large distances apply unchanged to a nonsymmetrical slender body.

For an axisymmetric slender body, it is assumed that the velocities in the respective coordinate directions are

$$v_r = v_r', v_x = U_\infty + v_x' \tag{7.21}$$

and the *velocity potential* is

$$\phi = U_\infty x + \phi'. \tag{7.22}$$

The perturbed quantities are quantities perturbed from infinite approaching flow because of the presence of a body.

The velocities and the velocity potential are related by the equations

$$v_x' = \frac{\partial \phi'}{\partial x}, v_r' = \frac{\partial \phi'}{\partial r}. \tag{7.23}$$

In a steady flow with respect to a flying aircraft, the equation to describe the flow, under the assumption of small perturbations for slender bodies, is given by the relation

$$-(M_\infty^2 - 1)\frac{\partial(rv_x')}{\partial x} + \frac{\partial(rv_r')}{\partial r} = 0,$$ (7.24)

which in terms of velocity potential is

$$-(M_\infty^2 - 1)\frac{\partial^2 \phi'}{\partial x^2} + \frac{1}{r}\frac{\partial}{\partial r}\left(r\frac{\partial \phi'}{\partial r}\right) = 0.$$ (7.25)

Equation (7.24) is a hyperbolic equation, unlike the flow equation in a subsonic case, which is elliptic in nature. The two equations differ considerably, as do their solutions.

For a point source situated at $x = 0, r = 0$ in an approaching flow, the perturbed velocity potential ϕ' in a low subsonic flow must be a function of radial distance only. For supersonic flows, this is slightly modified, and thus in cylindrical coordinates the expression for ϕ' is

$$\phi' = -\frac{q}{4\pi\sqrt{x^2 - (M_\infty^2 - 1)r^2}},$$ (7.26)

where q is the *source strength*. It is noted further that the two second derivatives are given by

$$\frac{1}{r}\frac{\partial}{\partial r}\left(r\frac{\partial \phi'}{\partial r}\right) = \frac{q(M_\infty^2 - 1)}{4\pi}\left[\frac{2}{(x^2 - (M_\infty^2 - 1)r^2)^{3/2}} + \frac{3(M_\infty^2 - 1)r^2}{(x^2 - (M_\infty^2 - 1)r^2)^{5/2}}\right]$$ (7.27)

and

$$\frac{\partial^2 \phi'}{\partial x^2} = -\frac{q}{4\pi}\left[\frac{1}{(x^2 - (M_\infty^2 - 1)r^2)^{3/2}} + \frac{3x^2}{(x^2 - (M_\infty^2 - 1)r^2)^{5/2}}\right],$$ (7.28)

and thus the differential equation is satisfied.

The field of a supersonic flow over a body of revolution at zero angle of attack is obtained by putting a source distribution along the axis of the body *source distribution*, $f(\xi)$, from $\xi = 0$ to $x - r\sqrt{M_\infty^2 - 1}$. If $\xi > x - r\sqrt{M_\infty^2 - 1}$, then the contribution of ϕ' is imaginary and the influence of the source distribution does not reach the point (x, r). Thus the expression for the perturbed velocity potential is

$$\phi' = \int_0^{x - r\sqrt{M_\infty^2 - 1}} \frac{f(\xi)}{\sqrt{(x - \xi)^2 - r^2(M_\infty^2 - 1)}}d\xi,$$ (7.29)

where

$$f(\xi) = \frac{1}{2}\frac{d(R^2)}{d\xi} - R\frac{dR}{d\xi} = \frac{A_o}{2\pi} \tag{7.30}$$

in which $R = R(x)$ is the cross-section radius and A_o the cross-sectional area of the axisymmetric body.

Evaluation of integral (7.29) at the upper boundary condition poses problems since the denominator becomes zero. Following the usual practice for supersonic flows, therefore, the following transformations are made:

$$\xi = x - r\sqrt{M_\infty^2 - 1}\cosh\sigma,$$

$$d\xi = -r\sqrt{M_\infty^2 - 1}\sinh\sigma d\sigma = -r\sqrt{M_\infty^2 - 1}\left[x - r\sqrt{M_\infty^2 - 1}\cosh\sigma\right]^{0.5}d\sigma$$

$$= \left[(x-\xi)^2 - r^2(M_\infty^2 - 1)\right]d\sigma,$$

and thus

$$\phi'(x,r) = \int f d\sigma. \tag{7.31}$$

Further, one gets the following relations for velocity components:

$$v_x' = \frac{\partial\varphi'}{\partial x} = \int f' d\sigma = \int_0^{x-r\sqrt{M_\infty^2-1}} \frac{f(\xi)}{\sqrt{(x-\xi)^2 - r^2(M_\infty^2 - 1)}}d\xi, \tag{7.32}$$

$$v_r' = \frac{\partial\varphi'}{\partial x} = -\int \sqrt{(M_\infty^2 - 1)}f'\cosh\sigma d\sigma, \tag{7.33}$$

$$= -\frac{1}{r}\int_0^{x-r\sqrt{M_\infty^2-1}} \frac{f(\xi)(x-\xi)}{\sqrt{(x-\xi)^2 - r^2(M_\infty^2 - 1)}}d\xi. \tag{7.34}$$

The method by which the linearized theory must be modified is based on the idea that the linearized solution may have the right form, but not quite in the right place. The remedy is to slightly strain the coordinates by expanding one of them as well as the dependent variables in asymptotic series. Thus the method is based on the hypothesis that the linearized theory gives a correct first approximation $\phi'(x_1,r)$ provided this is determined at (x,r), in which x_1 and r are axial coordinates on exact and approximate characteristics emanating from the same point on the body surface.

The equation for the characteristics that determines $y(x,r) = x - r\sqrt{M_\infty^2 - 1}$ is highly complicated and in general is not needed. For large values of r, Whitham's results [128] to determine perturbation velocities for a smooth body are as follows:

$$\frac{v'_x}{U_\infty} = -\frac{F(y)}{\sqrt{2\alpha r}},\tag{7.35}$$

$$\frac{v'_r}{U_\infty} = -\alpha v'_x,\tag{7.36}$$

where

$$F(y) = \int_0^\infty \frac{f(\xi)}{\sqrt{y-\xi}}, \alpha = \sqrt{M_\infty^2 - 1}.\tag{7.37}$$

Further, his equation for the characteristic of a real gas is

$$x = \alpha r - kF(y)\sqrt{r} + y,\tag{7.38}$$

where

$$k = \frac{(r+1)}{\sqrt{2}} \frac{M_\infty^4}{\alpha^{3/2}}.\tag{7.39}$$

The F curve gives immediately a rough description of the flow pattern since it shows whether the characteristics are converging in compression [$F'(y) > 0$], when a shock will appear, or is diverging in expansion [$F'(y) < 0$].

When the distance from the body is sufficiently large, the expression for pressure becomes, irrespective of whether the point is on a shock or not,

$$\frac{p - p_\infty}{p_\infty} = \frac{\gamma M_\infty^2 F(y) - \sqrt{2\alpha r}}{\sqrt{2\alpha r}}.\tag{7.40}$$

Discontinuities in pressure occur as shocks are crossed since the value of y specifying the characteristic on which the point (x, r) lies will change discontinuously. By a combination of the front and rear weak shocks, the pressure signature at a very large distance is found to be that of an *N-wave* and can be calculated from the preceding equation. The drag on the body can then be found either directly from the pressure distribution acting on the body or indirectly from the rate at which the energy or longitudinal momentum is transported across a *control surface* enclosing the body, which is given by *Whitham* as follows:

$$D = \pi\rho_\infty U_\infty^2 \int_0^\infty F^2(y)dy,\tag{7.41}$$

and thus the relation for the *drag coefficient* is

$$c_D = \frac{D}{(1/2)\rho_\infty U_\infty^2 A_{\text{ref}}} = \frac{2\pi}{A_{\text{ref}}} \int_0^\infty F^2(y)\, dy.\tag{7.42}$$

The preceding two expressions for drag contain more than drag on a moving body alone, which can be found by having a wake of uniform cross section equal to the base, and the difference is interpreted by *Whitham* as the contribution of the base pressure and, thus, the base drag in the wake-free region.

Although Whitham's analysis helped in the prediction and explanation of an *N-Wave* of the pressure signature in a satisfactory manner, the examples are valid for a homogeneous atmosphere only. The theory has, however, been extended by making allowances for nonlinear effects by Rao [94] and further by Randall [93], who investigated the effect on sonic bangs in which the phenomena occur. Finally, a detailed review of all these theories has been given, including the effect of lift (singularities of droplets), by Warren and Randall [127].

The method to predict the pressure signature for a supersonic flying aircraft, described previously, is highly complicated and need not be used if only the *cutoff Mach number* of the sonic boom for an aircraft flying at different altitudes is needed.

Let there be on ground a boom or bang from a supersonic flying aircraft if at any previous time the component of the velocity of the aircraft toward the point is at least equal to the speed of sound. In general, at an instant t, a point receives a wavelet generated by the aircraft at some single instant τ, but under certain conditions it may receive wavelets generated at successive instants to cause a sudden increase in pressure at the point. Thus a bang is heard if

$$t = \tau + r/c_o = \tau + \mathrm{d}\tau + (r + \mathrm{d}r)/c_o, \tag{7.43}$$

and thus

$$\frac{\mathrm{d}r}{\mathrm{d}\tau} = -c_o. \tag{7.44}$$

The quantity c_o is the speed of sound in a uniform atmosphere at rest. Consider the general case of a climbing aircraft in which

$$
\begin{aligned}
V_a &= \text{speed of the aircraft,} \\
c_a &= \text{speed of sound at the altitude of the aircraft,} \\
\gamma &= \text{climb angle,} \\
V &= \text{component of } V_a \text{ parallel to the ground} = c_g, \\
c_g &= \text{ speed of sound on the ground,} \\
M_a &= V_a/c_a.
\end{aligned}
$$

Noting the relations

$$\sin \mu = c_a/V = c_a/c_g, \mu_a = \mu - \gamma, V_a \sin \mu_a = V \sin \mu = c_g \sin \mu \tag{7.45}$$

the cutoff Mach number for the sonic boom to be heard is

$$\frac{V_a}{c_a} = M_a = \frac{c_g \sin \mu}{c_s \sin(\mu - \gamma)} = \frac{c_g/c_a}{\cos \gamma - \sin \gamma \sqrt{(c_g/c_a)^2 - 1}}. \tag{7.46}$$

Numerical results of (7.46) for different flight altitudes and climb rates are given in Table 7.1. These results show that in level flight and at moderate supersonic speeds the shock generated at the nose of an aircraft can't be avoided. However, detailed calculation of the shock structure by the method outlined earlier may show the possibility of weakening or eliminating the sonic boom altogether.

Table 7.1 Cutoff Mach numbers at different flight altitudes

		Cutoff Mach number M_a for climb angle $\gamma =$				
R(km)	c_g/c_a	$-20°$	$-10°$	$0°$	$+10°$	$+20°$
0	1.000	1.065	1.015	1.000	1.015	1.065
3	1.035	1.005	1.002	1.035	1.150	1.220
5	1.060	1.002	1.011	1.060	1.150	1.225
8	1.105	1.005	1.035	1.105	1.228	1.341
10	1.125	1.010	1.045	1.125	1.260	1.390
15	1.150	1.013	1.060	1.150	1.302	1.545
20	1.150	1.013	1.060	1.150	1.302	1.545

7.5 Measurement Techniques

While different mechanisms of noise production were discussed in earlier chapters, it may be worthwhile to look into some of the experimental measurement techniques for noise and also the reduction of noise. Some of the experimental setups for the measurement of jet noise have been described by Krishnappa and Csanady [55] and by Mollo-Christensen [78]. A schematic sketch of the experimental setup used by *Mollo-Christensen* is shown in Fig. 7.2.

This fairly old description of an experimental setup may have been superseded by the latest measurement techniques, but it will still be instructive to discuss this setup. In this method, while the sound level is measured with a calibrated microphone, the turbulence level is measured with a calibrated hot wire anemometer, from which the data are reduced by a setup schematically shown in Fig. 7.3 for correlation measurement.

Different measuring instruments and quantities to be measured are as follows:

(a) Sound level – ear phones (spaced, single or double)

   – sound level meter using microphone and filter

(b) Sound spectrum – recording sound spectrograph
(c) Structural vibration – magnetic moving coil

   – Piezoelectric pickup

(d) Frequency – quartz crystal oscillator, cathode ray oscilloscope
(e) Pressure amplitudes – absolute pressure measurements

   – Comparison with standard microphone
   – Hot wire method
   – Piston microphones

For the *calibration* of the aforementioned instruments, the following standard instruments are available:

(a) Frequency – quartz crystal oscillator (typically 10^6 (s^{-1}))

   – Precision fork (at low frequencies)
   – RC network (not very stable)

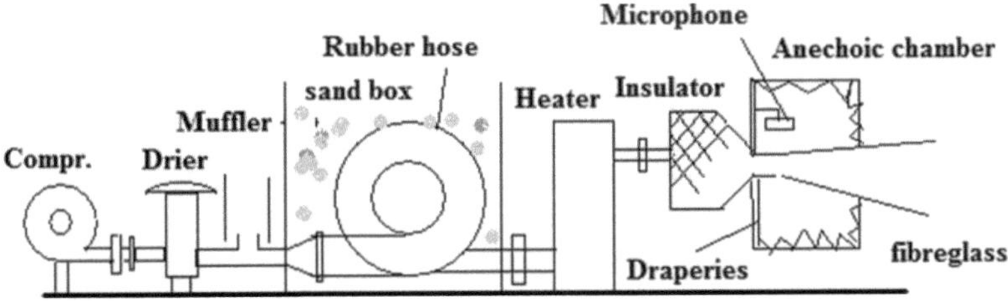

Fig. 7.2 Experimental setup for jet noise measurement

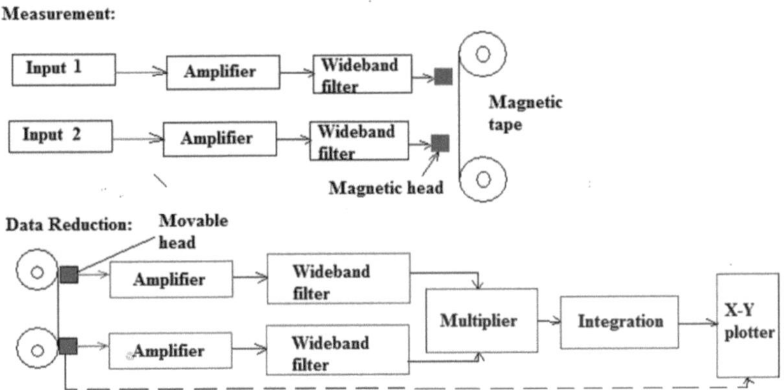

Fig. 7.3 Schematic sketch for measurement and data reduction for correlation

(b) Frequency by comparison – beat methods

-   Cathode ray oscilloscope
-   Stroboscope
-   Electronic meters (accuracy $\pm$ 1)
-   Vibrating rods

(c) Sound spectrum – piston phone

Several analog and digital *sound level meters* are commercially available in which the former are very cheap, with several decibel ranges. However, these are generally for broadband noise, and spectrum analysis is not possible.

The human ear is not equally responsive to all frequencies; it is most sensitive around 4,000 Hz and least sensitive in low frequencies. Responses to the sound level are modified by frequency-weighting networks that approximate the equal loudness-weighting networks or scales (*multiple scales*); some also have other scales like A, B, or C scales. The A scale, which approximates the ear's response to moderate sound, is commonly used in measuring noise to evaluate its effect on humans and has been incorporated into many occupational noise standards. In this connection, we should mention the various types of *airport noise*, which include *noise pollution* produced by any aircraft or its components during various phases of flight: on the ground while parked in connection with auxiliary power units, while taxiing during runup from propellers and jet exhaust, during takeoff, during arrival, overflying while en route, or during landing.

The shape of the nose, windshield, or canopy of an aircraft affects the sound produced. Much of the noise of a propeller aircraft is due to flow around the blades; a helicopter's main and tail rotors also give rise to noises that are mostly of low frequency and determined by the rotor speed.

Typically, noise is generated when flow passes an object on the aircraft, for example the wings or landing gear. There are, broadly speaking, two main types of *airframe noise*:

(a) *Bluff-body noise* due to alternate vortex shedding from either side of a bluff-body, creating low-pressure regions at the core of the shed vortices, which manifest themselves as pressure or sound waves;
(b) *Edge noise* when the turbulent flow passes the end of an object or gaps in a structure.

In order to control *airport noise* the U.S. Congress declared in 1972 its policy to promote an environment for all Americans free from noise that jeopardizes their health or welfare.

As a consequence, the Federal Aviation Administration (FAA) has the authority to regulate various questions related to flying, including airport noise. New FAA regulations published recently define the airspace under control of the tower to increase the ability to limit noise from airplanes and helicopters, by which the controllers would be able to keep arriving aircraft at a higher altitude longer or departing planes may be directed to climb more steeply and direct their flight paths away from developed areas. As a result of these regulations, most major airports do not allow takeoffs and landings outside of certain hours.

7.6   Reduction and Optimization Problems of Noise

While earlier discussions pertained to the production of noise, it is now possible to enumerate the ways by which noise can be reduced: (1) at the source, (2) in the manner of transmission by absorbing materials, (3) at the point of observation by changing the direction of propagation of the source, and (4) at the point of observation by changing the frequency.

For subsonic jet noise the sound pressure level at the source can be reduced for same thrust by (1) reducing the average jet velocity in a bypass engine, (2) reducing the jet diameter by totally or partially partitioning (multiple lobes and tubes), and (3) reducing the shear stress and consequently the turbulence level in the mixing region (by entrainment of gas, blowing of gas, or by reduction of the primary turbulence level by screens). A point for bypass engines to be made incidentally at this point is about higher *propulsive efficiency* of such engines for aircraft flying in the transonic range. It is well known that the maximum propulsive efficiency for

Fig. 7.4 Multiple tube and lobe nozzles

a jet engine occurs when the jet velocity equals the flight velocity of an aircraft. Thus for subsonic and transonic flying aircraft the jet velocity may be reduced as much as possible but it may still be larger than the flight velocity of the aircraft to develop the required thrust. For engines in Boeing 707 aircraft, the fuel consumption for a bypass engine, when the engine was first introduced, was reported to be approximately 15% lower than the equivalent straight jet engine it replaced with the same thrust, in addition to having the superior noise characteristics of a bypass engine (or a fanjet engine with a very high bypass fan air to core air mass flow ratio) over a straight jet engine. Further, for jet engines considerable reduction of sound pressure level was achieved by introducing multiple tube or multiple lobe nozzles (Fig. 7.4).

Reduction of noise from fans was achieved also by putting insulation materials in the inlet, by increasing the spacing between the rotor and the stator, and by changing the number of leading edge slope of the blades (change in acoustic frequency!). Further improvements have been reported by mounting the engine above the wing instead of below the wing to reduce the engine noise in flight (in stealth aircraft; however, the engine may be put above the wing to reduce or eliminate the radar signature from the ground).

A jet engine noise reduction device, called a *chevron nozzle*, is now in use on commercial airliners and is a good example of a *NASA developed technology* that reflects years of paper studies, prototype building, and testing. Chevrons are the saw-tooth pattern that can be seen on the trailing edges of some jet engine nozzles. As hot air from the engine core nozzle mixes with cooler air blowing through the engine fan, the jagged nozzle serves to smooth the mixing, which reduces the turbulence that causes a reduction in noise.

Uzun and Hussainy [125] and Schlinker et al. [99] have studied the reduction of noise in chevron subsonic and supersonic nozzles, respectively. In the *Uzun and Hussainy* paper, simulation of the near-nozzle region of a moderate Reynolds number cold jet flow exhaust from a *chevron nozzle* was reported, for which the SMC001 nozzle, experimentally studied by researchers at the *NASA Glen Research Center* earlier, was used. "The nozzle angle design contains six symmetric chevrons that have a five-degree penetration angle. The flow inside the chevron nozzle and the free-jet flow outside are computed simultaneously by a high-order accurate, multi-block, large-eddy simulation (LES) code due to chevrons with overset grid capability. The total number of grid points at which the governing equations are solved is about 100 million. The main emphasis of the simulation is to capture the enhanced shear-layer mixing, and consequent (reduction) noise generation, that occurs in the mixing layer of the jet within the first few diameters downstream of the nozzle exit."

In this study, the Favre-filtered, unsteady, compressible, nondimensionsional Navier–Stokes equations in curvilinear coordinates in conservative form are written as

$$\frac{\partial \mathbf{Q}}{\partial t} = -\left[\frac{\partial}{\partial \xi}\left(\frac{\mathbf{F}-\mathbf{F_v}}{J}\right) + \frac{\partial}{\partial \eta}\left(\frac{\mathbf{G}-\mathbf{G_v}}{J}\right) + \frac{\partial}{\partial \zeta}\left(\frac{\mathbf{H}-\mathbf{H_v}}{J}\right)\right] = 0, \quad (7.47)$$

where t is the time, ξ, η, and ζ are the *general curvilinear coordinates* of the computational space, J is the *Jacobian of the coordinate transformation* [1,2] from the physical domain to the public domain, $\mathbf{Q} = Q/J$, where $Q = [\bar{\rho}, \bar{\rho}\bar{u}, \bar{\rho}\bar{v}, \bar{\rho}\bar{w}, \bar{e}_t]^{\mathrm{T}}$ is the *vector of the conservative flow variables*, F and G are *inviscid flux vectors*, and $\mathbf{F_v}$, $\mathbf{G_v}$, and $\mathbf{H_v}$ are the *viscous flux vectors*.

The viscous stress terms in the governing equations are obtained using the first derivatives of the velocity components, and the second derivatives of the viscous terms are evaluated by the application of the first derivative operator.

The LES code employs the following tridiagonal spatial filter, except near a boundary, where special arrangents have to be made:

$$\alpha_f \bar{f}_{i-1} + \bar{f}_i + \alpha_f \bar{f}_{i+1} = \sum_{n=0}^{3} \frac{2a_n}{3}(f_{i+n} + f_{i-n}), \quad (7.48)$$

where $\bar{f}_i$ denotes the filtered value of the quantity f at grid point i and

$$a_0 = \frac{11}{16} + \frac{5\alpha_f}{8}, a_1 = \frac{15}{32} + \frac{17\alpha_f}{16}, a_2 = \frac{-3}{16} + \frac{3\alpha_f}{8}, a_3 = \frac{1}{32} - \frac{\alpha_f}{16}. \quad (7.49)$$

The parameter α_f must satisfy the inequality $-0.5 < \alpha_f < 0.5$. The time advancement can be performed by means of either the standard fourth-order explicit Runge–Kutta scheme or a second-order Beam-Warming type scheme.

Simulation of results from a cold jet from the SMC001 chevron nozzle has been carried out in which six chevrons penetrate the jet core flow by five degrees. The acoustic Mach number of the jet centerline exit velocity is 0.9, and the *Reynolds number* based on the jet nozzle centerline velocity and nozzle exit diameter is set to 100,000 in the simulation. Multiblock discretization is performed. A total of 512 parallel processors were used for computations on the 100 million grid points during a total run time of 12 days on a Cray XT3 machine.

A comparison of the noise spectra in the sideline direction showed that the spectrum obtained using the 100 million grid point captured the peak region of the spectrum very accurately.

Schlinker et al. [99] have studied supersonic jet noise from round and chevron nozzles experimentally. Their paper deals with laboratory scale jet noise experiments for a Mach number of $M_{\mathrm{jet}} = 1.5$ with a stagnation temperature ratio ranging from 0.75 to 2.0. The baseline configuration was represented by a round converging-diverging (CD) ideal expansion nozzle. A round CD nozzle with chevron was included as the first of several planned noncircular geometries directed at demonstrating the impact on large-scale structure noise and validating noise prediction methods for geometries of future technological interest. Overexpanded and underexpanded nozzles were also tested on both nozzle configurations, and the paper examines far-field spectra, directivity patterns, and overall sound pressure level dependence to compare the fine-scale turbulence noise and large-scale turbulence structure noise observed by other authors.

Historically, the first patent for a noise control system was granted to *Paul Lueg*; the patent described how to cancel sinusoidal tones in ducts by *phase advancing* the wave and cancelling arbitrary sounds in the region around a loudspeaker by inverting the polarity. In the 1950s, *Lawrence J. Fogel* received a U.S. patent to design systems to cancel the noise in helicopter and airplane cockpits. In 1957 he developed a working model of *active noise control* applied to an earmuff to have an active attenuation bandwith of approximately 50–500 Hz, with a maximum attenuation of approximately 20 dB. In 1986, *Dick Rutan* and *Jeana Yeager* used prototype headsets for a *Bose line of headphones* for their *around-the-world flight*. Hammond et al. [47] wrote a report on noise reduction efforts for special operations of C-130 aircraft using active control. *Aviation Week* journal in its issue of December 27, 2005 published a fascinating article on the design of *passive and active noise cancellation* inside the passenger cabin for high frequencies and active cancellation for low frequencies. The technology proposed was the same technology that is used in the *Bose line of headphones* where an out-of-phase signal is induced under the ear cup that cancels out the sounds that leak in from outside. For aircraft applications, the design tradeoff is governed by the constraints that an active cancellation does not work very well at high frequencies but does well at low frequencies, while the passive cancellation requires massive sound-absorbing materials. Thus the design tradeoff works out very neatly – active at low frequencies and passive at high frequencies. A proposal for this is to use a large number of microphones, say 24, to provide a cancellation algorithm to continuously adapt to the changing noise environment through a lower number of loudspeakers, say 12.

One of most recent efforts in the area of noise reduction technology was the *Quiet Technology Demonstrator* (QTD) program of *Boeing Company* conducted in partnership with *Rolls-Royce* using a newly built *Boeing 777 aircraft*. Later the program was extended in 2005 in cooperation with several engine manufacturers, NASA, and some other airliners, by including chevrons on the engine exhaust ducts, new acoustic treatment for the engine inlet, and landing gear noise reduction features with the goal of lower noise for both communities surrounding airports and also for passengers and crews in the cabin.

Noise heard on the ground is composed of engine noise and airframe noise. During takeoff, engine noise and airframe noise are important. During takeoffs, engine noise is dominant and airframe noise is not as important, because at that stage the landing gear is stowed and the flaps are only deployed to a small deflection. On the other hand, during landing, both engine noise and airframe noise are important, while on most modern airplanes, the generated noise is still a bit more than the airframe noise. During landing the *reverse jet* and brakes to reduce the landing speed of aircraft can also be considerable, although for a short time.

While the noise in a straight jet or fanjet happens due to a mixing of air streams to produce turbulent flow and noise, the fans within engines makes noise from both the front and back of the engine, consisting of both *broadband noise* and *discrete-frequency noise*, the latter due to noise from wakes of succeeding rows of rotor and stator blades. The fan noise can be controlled by engine design features and by including acoustically absorbent surfaces in the engine inlet nacelle. There can also be *buzz-saw noise*, which is a particular fan noise source that occurs when the tips of the fan blade travel close to the speed of sound, especially during takeoff.

The inlet nacelle surface is usually made of an acoustically absorbent material, so-called *acoustic lining*, consisting of a perforated sheet at the surface, behind the surface a layer of honeycomb material, and finally a nonporous back sheet. The depth of the honeycomb core and the properties of the perforated sheet (hole diameter, sheet thickness, core depth, etc.) determine which frequency of sound is absorbed (*absorbing frequency*). As a rule of thumb, the deeper the liner, the lower the frequency of sound that is absorbed, and the effectiveness of the acoustic lining extends to a small frequency range around the *absorbing frequency*. In order to make the frequency over the sound is absorbed wider, typical engine inlet liners have an additional perforated sheet in between the face sheet and the back sheet (*double layer liner*). There can also be foam material for acoustic lining.

Next, we discuss the reduction of *blade-cutting noise* in the turbomachinery. According to a 1978 patent issued to *Stephen B. Kazin* and *Ram K. Mehta* of GE, "A turbo-machinery stage consisting of two-axially spaced blade-rows, one of which is rotable, wherein the blades of the upstream row are contoured to present a leaned wake to the downstream row. In the preferred configuration, the blades of the upstream row extend radially from a central hub and at a radius above the hub, are physically curved circumstantially from a radial line above the hub. The physical lean locally complements the inherent aerodynamic lean at each radius such that the integrated acoustic power of the stage is essentially at a minimum, thus maximizing the acoustic attenuation. Preferably, the constraint is imposed that the blade physical curvature is essentially smooth, continuous and without an inflection point."

In a 1998 report by *Scott Sawyer* and *Sanford Fleeter* of Purdue University entitled *Passive Control of Turbomachine Noise*, the authors state: "Discrete-frequency tones generated by unsteady blade row interactions are of particular concern in turbo-machinery design. In the annular inlet and exit ducting, rotor–stator interactions generate acoustic waves at the multiple of rotor blade frequency. This rotor–stator generated discrete-frequency noise is characterized as a summation of propagating acoustic waves over the multitudes of the rotor blade pass frequency. Aerodynamic detuning is accomplished by the replacement of alternate stator vanes with short chord splitter vanes. The tunes stator vane row influences the unsteady aerodynamics and acoustic response of the rotor–stator interaction. The unsteady aerodynamics and acoustic response of detuned vane row are modeled analytically as a 2D flat plate cascade operating in inviscid compressible subsonic flow with small unsteady perturbations. The linearized continuity and momentum equations are solved using wave theory. The model is applied to the interaction of (single stage axial compressor) 16 bladed rotor and a 36 vaned stator with a reduced frequency of 8.0. The detuned stator vane row incorporates 36 half-chord splitters with 36 full-chord airfoils. The optimum configuration was determined for the detained rotor: offset 0.3 chord, spacing ratio 0.3, detuned pitch spacing 1.7 and reduced frequency 5.2. The tuned and detuned stator vane rows were modeled over a range of operating conditions corresponding to a range of Mach numbers from 0.09 to 0.4. Maximum reduction of 8 dB were realized, and aerodynamic detuning was effective over nearly the entire range of operating conditions."

Noise reduction in a supersonic inlet has been a topic of study by various investigators. *Christopher A. Sanders* in his M.S. thesis from Virginia Polytechnic Institute of 1998 studied noise reduction in an axisymmetric supersonic inlet using trailing edge blowing. "Acoustic experiments were conducted in an anechoic chamber with a 1/14th scale model of a supersonic aircraft engine inlet using *trailing edge blowing* (TEB) to reduce the engine fan noise from a *turbofan propulsion simulator* (TPS). The TPS is 10.4 cm in diameter and is powered by compressed air. The supersonic inlet is connected to the TPS and is geometrically and acoustically scaled from a working design. The supersonic inlet is operated in a take-off or landing operating condition where the inlet core flow is subsonic. TEB is the process of ejecting high pressure air to re-energize the wakes of upstream fan disturbances such as struts or inlet guide vanes (IGV). The elimination of the wakes will provide a uniform flow field at the engine fan face and reduce noise at the blade passing frequency. The TEB was implemented in six non-uniformly spaced support struts in the inlet. Acoustic tests were then performed at 40%, 60% and 88% of the fan design speed (PNC) to measure the reduction in the blade passing tone (BPT) due to TEB from the struts with and without the presence of IGV. The noise reductions without IGV at 40 PNC show the best results with the blade passing tone (BPT) being reduced by an average of 3.1 dB. The first harmonic of the BPT and the overall sound pressure level (SPL) were also reduced by 1 dB. The addition of the IGV in the inlet reduced the effectiveness of the TEB. The addition of IGV changed the reduction in BPT at 40 PNC by 0.5 dB and the overall SPL was unchanged. At 60 PNC the addition of IGV reduced the reduction due to TEB in the BPT from an average 2 dB to an average of 1 dB. The tests performed at 88 PNC showed negligible effects due to TEB. Aerodynamic experiments performed on the inlet that showed that the wakes of the IGV have a larger velocity defect than the struts, thus making the IGV a greater noise source."

Serrated trailing edge noise has been a topic of study by several researchers. Howe [51] investigated the aerodynamic noise of a serrated trailing edge. "A discussion is given of the production of sound by low Mach number turbulent flow over the trailing edge of a serrated airfoil. The airfoil is modeled by a flat plate set at zero angle of attack to the mean flow, and attention is given to both limiting cases in which the chord of the airfoil is either large or small relative to the characteristic acoustic wavelength; a formula is proposed for interpolating predictions at intermediate frequencies. General arguments are advanced which imply that, for serrations of spanwise wavelength λ and amplitude h, and at radian frequencies ω satisfying the relation $\omega h/U?h/U \gg 1$ (U being the velocity of the main stream), the frequency spectrum of trailing edge noise is reduced relative to that for an unserrated edge by $10 \times \log_{10}(C/\lambda$ (dB) when $\lambda/h \leq 4$, where the constant $C \simeq 10$. This conclusion is confirmed, and are extended to include larger values of λ/h, by an approximate analytical treatment of the case involving an edge with sinusoidal serrations, and by use of an empirical model of the turbulent flow. At high frequencies the predicted attenuation is about 1 dB when $\lambda/h = 10$; at $\lambda/h = 1$ the attenuation is typically 7 or 8 dB. It is argued that, in principle, optimal attenuations should be obtained by use of serrations of sawtooth profile with edges inclined at less than 45° to the direction of the mean flow.

Further, Geiger [42] of *Virginia Polytechnic Institute and State University, Blacksburg/VA* performed a comparative analysis of serrated trailing edge designs on idealized aircraft engine fan blades for noise reduction. "The effects of serrated trailing edge designs, designed for noise reduction, on the flow-field downstream of an idealized aircraft engine blade row were investigated in detail. The measurements were performed in the Virginia Tech low speed linear cascade tunnel on one set of baseline GE-Rotor-B blades and four sets of GE-Rotor-B blades with serrated trailing edges. The four serrated blade sets consisted of two different serration sizes (1.27 cm and 2.54 cm) and for each different serration size a second set of blades with added trailing edge camber. The cascade consisted of 8 GE-Rotor-B blades and 7 passages with adjustable tip gap settings. It had an inlet angle of 65.1°, stagger angle of 56.9° and a turning angle of 11.8°. The tunnel was operated with a tip gap setting of 1.65% chord, with a Reynolds number based on the chord of 390,000.

Blade loading measurements performed on each set of blades showed that it was slightly dependent on the serration shape. As the serration size was increased the blade loading decreased, but adding droop increased the blade loading.

The Pitot-static cross sections showed that flow fields near the upper and lower endwalls' cascade tunnel were similar with the baseline or the serrated blade downstream of the blade row. As the wake convected downstream, the individual tips and valleys could be seen. The tip leakage vortex was only minimally affected by the trailing edge serrations. This conclusion was further reinforced by the three-component hot-wire cross-sectional measurements that were performed from the lower endwall to the midspan of the blade. These showed that the mean streamwise velocity, turbulence kinetic energy, and turbulence kinetic energy production in the tip leakage region were nearly the same for all four serrated blades as well as the baseline. The vorticity in this region was more dependent on the serration shape and as a result increased with serration size compared to the baseline.

Midspan measurements performed with the three-component hot wire showed the spreading rate of the wake, and the decay rate of the wake centerline velocity deficit increased with serration size compared to the baseline case. Drooping of the trailing edge only minimally improved the spreading and decay rates. The improvement in these rates was predicted to reduce the tonal noise at the leading edge of the downstream stator vane because the periodic fluctuation associated with the sweeping of the rotor blade wakes across it was due to the pitchwise variation of the mean streamwise velocity. The wakes were further compared to the mean velocity and turbulence profiles of plane wakes, with which the baseline and the smallest serration size agreed best. As the serration size was increased and drooping was added, the wakes became less like plane wakes. Spectral plots at the wake centerline in all three velocity directions showed some evidence of coherent motion in the wake as a result of vortex shedding.

Shape optimization for noise control has been reported by Marsden et al. [74] of the University of Montpellier (France), for which the "noise generated by turbulent boundary layers near the trailing edge of the lifting surface continues to pose a challenge for many applications."

"For trailing-edge noise control, a shape design method based on control theory for partial differential equations and a gradient-based minimization algorithm is employed to optimize the trailing edge shape, [74]" although the main difficulty in the gradient-based optimization is the calculation of the gradient of the cost function with respect to the control parameters, and thus, the most widely used method is to solve an adjoint equation in addition to the flow solutions.

7.7 Exercises

7.7.1 What are the basic differences between *supersonic jet noise* and *subsonic jet noise*?

7.7.2 What are the different regions of a fully expanded supersonic jet ?

7.7.3 What is a *screeching noise*?

7.7.4 How is combustion noise generated?

7.7.5 What is the characteristic sonic-boom signature of a supersonic flying aircraft and how is it generated?

7.7.6 What are the cutoff Mach numbers for a sonic boom?

7.7.7 What is the difference between an *anechoic chamber* and a *reverberation chamber*?

7.7.8 How is the exit nozzle of a subsonic jet designed so as to reduce the external jet noise?

Erratum to:

Chapter 3
Lighthill's Theory of Aerodynamic Noise

T. Bose, *Aerodynamic Noise: An Introduction for Physicists and Engineers*, Springer
Aerospace Technology 7, DOI 10.1007/978-1-4614-5019-1,
© Springer Science + Business Media New York 2013

DOI 10.1007/978-1-4614-5019-1_8

Equation 3.2 should appear as follows:

$$\frac{\partial(\rho u_i)}{\partial t} + \sum_j \frac{\partial(\rho u_i u_j)}{\partial x_j} = \sum_j \frac{\partial \tau_{ij}}{\partial x_j} + F_i, \mathrm{Nm}^{-3}, \tag{3.2}$$

Equation 3.3 should appear as follows:

$$\tau_{ij} = \left(-p - \frac{2}{3}\mu \sum \frac{\partial u_j}{\partial x_j} \right) \delta_{ij} + \mu \left(\frac{\partial u_i}{\partial x_j} + \frac{\partial u_j}{\partial x_i} \right) = -p\delta_{ij} - \tau_{ij}^*,$$

$$\delta_{ij} = \text{Kronecker delta: } \delta_{ii} = 1, \delta_{ij} = 0,$$

$$\tau_{ij}^* = \left(\frac{2}{3}\mu \sum \frac{\partial u_j}{\partial x_j} \right) \delta_{ij} - \mu \left(\frac{\partial u_i}{\partial x_j} + \frac{\partial u_j}{\partial x_i} \right). \tag{3.3}$$

Equation 3.4 should appear as follows:

$$\frac{\partial(\rho u_i)}{\partial t} + \sum_j \frac{\partial(\rho u_i u_j)}{\partial x_j} = -\frac{\partial p}{\partial x_j} - \sum_j \frac{\partial \tau_{ij}^*}{\partial x_j} + F_i, \mathrm{Nm}^{-3}. \tag{3.4}$$

Equation 3.6 should appear as follows:

$$\sum_i \frac{\partial^2 p}{\partial x_i^2} = -\sum_i \frac{\partial^2(\rho u_i)}{\partial t \partial x_i} - \sum_i \sum_j \frac{\partial(\rho u_i u_j)}{\partial x_i \partial x_j} - \sum_i \sum_j \frac{\partial^2 \tau_{ij}^*}{\partial x_i \partial x_j} + \sum_i \frac{\partial F_i}{\partial x_i}. \tag{3.6}$$

Equation 3.10 should be replaced by the following text:
Strictly speaking, the second term in the above equation should have T_{ij}^*, but here it
has been replaced by T_{ij}, because Lighthill considers quadruple effect only through
the convective term and through the fluctuation of stress terms.

The online version of the original chapter can be found at
http://dx.doi.org/10.1007/978-1-4614-5019-1_3

References

1. Bose, T.K.: Computational Fluiddynamics, 1st Reprint. Eastern Wiley, New Delhi (1990)
2. Bose, T.K.: Numerical Fluid Dynamics. NAROSA, Delhi (1997)
3. Goldstein, M.: Aeroacoustics. McGraw-Hill, New York (1976)
4. Harris, C.M. (ed.): Handbook of Noise Control. McGraw-Hill, New York (1957)
5. Morse, P.M., Feshbach, H.: Methods of Theoretical Physics, Part 1 (Chapters 1 to 8). McGraw-Hill, New York (1953)
6. Morino, L.: Mathematical foundations of integral methods. In: Morino, L. (ed.) Computational Methods in Potential Aerodynamics, pp. 271–291. Springer, Berlin (1985)
7. Rayleigh, L.: Theory of Sound, 2 Vols, 1895, (paperback edition, Dover; google edition: http://books.google.com)
8. Ribner, H.S.: The generation of sound by turbulent jets. Advances in Applied Mechanics, vol. III, p. 103. Academic, New York (1900)
9. Sagaut, P.: Large Eddy Simulation for Incompressible Flows: An Introduction, 3rd edn. Springer, Berlin (2006)
10. Speziale, C.G.: Computing non-equilibrium flows with time dependent RANS and VLES. 15th ICNMFD, Lecture Notes in Physics. Springer, Berlin (1996)
11. Stratton, J.A.: Electromagnetic Theory. McGraw-Hill, New York (1941)
12. Wagner, C., Huettl, T., Sagaut, P.: Large-Eddy Simulation for Acoustics. Cambridge University Press, Cambridge (2007)
13. Ahuja, V., Mendonca, Y., Long, L.N.: Computational simulation of fore and aft radiation from a ducted fan, 6th Aeroacoustics Conference, Lehaina/HI, June 12–14, 2000, AIAA Paper 2000-1943 (2000)
14. Alberts, D.J.: Addressing Wind Turbine Noise. Report from Lawrence Technological University (2006)
15. Anderson, N., Eriksson, L.-E., Davidson, L.: Large-eddy simulation of a Mach 0.75 Jet. 9th AIAA/CEAS Aeroacoustics Conference and Exhibit, AIAA 2003-3312, Hilton Head, SC, 12–14 May 2003
16. Arunajatesan, S., Sinha, N.: Unified unsteady RANS-LES simulations of cavity flow field. 39th AIAA Aerospace Sciences Meeting, 8–11 Jan. 2011, Reno/NV, AIAA Paper 2001-0516 (2006)
17. Avital, E.J., Sandham, N.D., Luo, K.H.: Sound generation closing data from numerical simulations of mixing layers, 2nd AIAA/CEAS Aeroacoustics Conference, May 6–8, 1996, State College/PA, AIAA Paper 96-1728 (1996)
18. Bailly, C., Juve, D.: Numerical solution of acoustic propagation problems using linearized Euler equations. AIAA J. **38**, 22 (2000)

T. Bose, *Aerodynamic Noise: An Introduction for Physicists and Engineers*,
Springer Aerospace Technology 7, DOI 10.1007/978-1-4614-5019-1,
© Springer Science+Business Media New York 2013

19. Bailly, C., Bogey, C., Juve, D.: Computation of flow noise using source terms in linearized Euler equations. 6th AIAA/CEAS Aeroacoustics Conference, AIAA 2000-2047, Lahaina/Hawaii, 12–14 June 2000; AIAA J. **42**(2) (2004)
20. Batten, P., Goldberg, U. and Chakravarthy, S.: Sub-grid turbulence modeling for unsteady flow with acoustic resonance, 38th AIAA Aerospace Sciences Meeting & Exhibit, Jan 10–13, 2000, AIAA Paper 2000-473 (2000)
21. Batten, P., Goldberg, U., Chakravarthy, S.: Interfacing statistical turbulence closures with large eddy simulation. AIAA J. **42**, 485 (2004)
22. Batten, P., Goldberg, U., Kang, E., Chakravarthy, S.: Smart sub-grid scale models for LES and hybrid RANS/LES, 6th AIAA Theoretical Fluid Mechanics Conference, 27–30 June 2011, Honolulu/HI, 2011-3472 (2011)
23. Bayliss, A., Maestrello, L., Parikh, P., Turkel, E.: A fourth order scheme for the unsteady compressible Navier–Stokes equations. AIAA Fluid dynamic and plasma dynamic Conference, Cincinnati, AIAA 85-1694, 16–18 July 1985
24. Boersma, B.J., Lele, S.K.: Large Eddy simulation of a Mach 0.9 turbulent jet. AIAA/CEAS Aeroacoustics Conference, AIAA Paper 1999-1874, Bellevue, WA, 10–16 May 1999
25. Boersma, B.J.: Large eddy simulation of the sound field of a round turbulent jet. Theor. Comput. Fluid Dyn. **19**, 161–170 (2005)
26. Bogey, C., Bailly, C., Juve, D.: Computation of the sound radiated by a 3-D jet using large eddy simulation. 6th AIAA/CEAS Aeroacoustics Conference, AIAA 2000–2009, Lahaina/Hawaii, 12–14 June 2000
27. Bogey, C., Bailly, C., Juve, D.: Noise investigation of a high subsonic moderate Reynolds number jet using a compressible LES. Theor. Comput. Fluid Dyn. **16**, 273–297 (2003)
28. Bose, T.K.: Flow properties fluctuations in a convergent-divergent nozzle. J. Spacecraft Rockets **10**, 530 (1973)
29. Carley, M.: Time domain calculation of noise generated by a propellar in a flow. Ph.D. Thesis, Dublin University (1996)
30. Chyczewski, T.S., Morris, P., Long, L.N.: Large-eddy simulation of wall bounded shear flow using the non-linear disturbance equations. 6th AIAA/CEAS Aeroacoustics Conference, Lahaina, HI, 12–14 June 2000
31. Chyczewski, T.S., Long, L.N.: Numerical prediction of the noise produced by a perfectly expanded rectangular jet, 2nd AIAA/CEAS Aeroacoustics Conference, May 6–8, 1996, State College/PA, AIAA Paper 96-1730 (1996)
32. Coltman, J.W.: Acoustics of the flute. Phys. Today **21**(11), 25–32 (1968)
33. Colonius, T., Lele, S.K., Moin, P.: Sound generation in mixing layer. J. Fluid Mech. **330**, 375–409 (1997)
34. Csanady, G.I.: The effects of mean velocity variation on jet noise. J. Fluid Mech. **26**, Part 1, 183–197 (1966)
35. Curle, N.: The influence of solid boundaries upon aerodynamic sound. Proc. R. Soc. **231A**, 505–514 (1955)
36. Davies. P.O.A.L., Fischer, M.J., Barrat, M.J.: The characteristics of the turbulence in the mixing region of a round jet. J. Fluid Mech. **15**, 337–367 (1963)
37. Freund, J.B.: Acoustic sources in a turbulent jet: a direct numerical simulation study. 5th AEAA/CEAS Aeroacoustics Conference, AIAA Paper 99-1858, Bellevue, WA, 10–12 May 1999
38. Freund, J.B.: Noise source at in low reynoldsnumber turbulent jet. J. Fluid Mech. **438**, 277–305 (2001)
39. Ffowcs-Williams, J.F.: The noise from turbulence convected at high speed flow. Phil. Transac. Royal Soc. **255A**, No. 1061, 469–503 (1963)
40. Frink, N.: Assessment of an unstructured grid method for predicting 3d turbulent viscous flow. 34th AIAA Aerospace Sciences Meeting and Exhibit, 15–18 Jan., 1996, Reno/NV, AIAA Paper 96-0292 (1996)

41. Garrison, L.A., Lyrintzis, A.S., Blaisdell, G.A., Dalton, W.N.: Computational fluiddynamics analysis of jets with internal forced mixures. 17th AIAA/CEAS Aeroacoustics Conference, AIAA Paper 2005-2887, Monterey/CA, 23–25 May 2005
42. Geiger, D.H.: Comparative analysis of serrated trailing edge designs on idealized aircraft engine fan blades for noise reduction. M.S. thesis, VPI & SU, Blacksburg, VA, 2004
43. George, A.R.: Helicopter noise – state of the art. J. Aircraft **15**, 707 (1978)
44. Goldberg, U., Batten, P., Palaniswamy, S., Chakravarty, S., Peroomian, O.: Hypersonic flow prediction using linear and non-linear turbulence closures. J. Aircraft **27**(4), 671–675 (2000)
45. Gupta, K.K., Choi, S.B., Ibrahim, A.: Aeroelastic acoustic simulation of flight systems. AIAA J. **48**(4), 798–807 (2010)
46. Gupta, K.K., Choi, S.B., Ibrahim, A.: Development of aerothermoelastic acoustics simulation capability of flight vehicles. 48th AIAA Aerospace Sciences Meeting, 4–7 Jan., 2010, Orlando/FL, AIAA Paper 2010-50 (2010)
47. Hammond, D., McKinley, R., Hale, B.: Noise reduction efforts for special operations C-130 aircraft using active synchrophaser control. Air Force Research Lab. Report ASC-99-0898 (1999)
48. Hixon, R.: 3D characteristic-based boundary conditions for CAA. 6th AIAA/CEAS Aeroacoustics Conference, AIAA 2000–2004, Lahaina, HI, 12–14 June 2000
49. Hixon, R., Shih, S.-H., Mankbadi, R.R.: Evaluation of boundary conditions for computational aeroacoustics. AIAA J. **33**(11), 2006–2012 (1995)
50. Hixon, R.: On increasing the accuracy of MaCormack schemes for aeroacoustic applications. AIAA/CEAS Aeroacoustics Conference, AIAA-97-1586-CP, Atlanta, GA, 12–14 May 1997
51. Howe, M.S.: Aerodynamic noise of a serrated trailing edge. J. Fluid Struct. **5**(1), 33–45 (1991)
52. Jones, W.P., Launder, B.E.: The prediction of laminarization with a two-equation model of turbulence. Int. J. Heat Mass Transfer **15**, 301–314 (1972)
53. Kennedy, C.A.: Low storage, explicit runge-kutta schemes for the compressible navier-stokes equations ICASE report No. 99–22, NASA/CR-1999-209349, NASA/Langley, Hampton/VA (1999)
54. Kolbe, R., Kailasnath, K., Young, T., Boris, J., Landsberg, A.: Numerical simulations of flow modification of supersonic rectangular jets. AIAA J. **34**(5), 902–908 (1996)
55. Krishnappa, G., Csanady, G.T.: An experimental investigation of the composition of jet noise. J. Fluid Mech. **37**, Part 1, 149–159 (1969)
56. Lassiter, L.W., Hubbard, H.H.: Experimental studies of noise from subsonic jets in still air. NASA TN-2757 (1952)
57. Lee, A., Harris, W.L., Widnall, S.E.: A study of helicopter rotated noise. J. Aircraft **14**, 1126 (1977)
58. Lew, P., Uzun, A., Blaisdell, G.A., Lyrintzis, A.S.: Effects of inflow forcing on jet noise using large eddy simulation. 45th Aerospace Scientific Conference, AIAA Paper 2004-0516, Reno/NV, 18-14 Jan 2004
59. Lew, P., Blaisdell, G.A., Lyrintzis, A.S.: Recent progress of hot jet aeroacoustics using 3-D large eddy simulation. AIAA Paper 2005-3084 (2005)
60. Lew, P.-T., Blaisdell, G.A., Lyrintzis, A.S.: Investigation of noise sources in turbulent hot jets using large eddy simulation data, 45th AIAAero Science Meeting, AIAA Paper 2007-16, Reno/NV, 8–11 January 2007
61. Lighthill, M.J.: On sound generated aerodynamically. I. General theory. Proc. R. Soc. Lond. **211A**, 564–587 (1952)
62. Lighthill, M.J.: On sound generated aerodynamically. II. Turbulence as a source of sound. Proc. R. Soc. Lond. **222A**, 1–32 (1954)
63. Lighthill, M.J.: Jet noise. AIAA J. **1**(7), 1507–1517 (1963)
64. Lloyd, P.: Some aspects of engine noise. J. R. Aeron. Soc. **63**, 541 (1959)
65. Louis, J.F., Patel, J.R.: A Systematic Study of Supersonic Jet Noise. Gas Turbine Lab., Massachusetts Institute of Technology, Cambridge (1971), GTI Report 107
66. Lyrintzis, A.S.: The use of Kirchhoff's method in computational aeroacoustics. ASME J. Fluids Eng. **116**, 665 (1994)

67. Lyrintzis, A.S.: Integral methods in computational aeroacoustics – from the (CFD) new field to the (Acoustic Far Field), CEAS Workshop, Athens/Greece, 2002

68. Mawardi, O.K.: On the spectrum of noise from turbulence. J. Acoust. Soc. Am. **27**(3), 442–445 (1955)

69. Mankbadi, R.R., Hayder, M.E., Povinelli, L.A.: The structure of supersonic jet flow and its radiated sound. 31st Aerospace Sciences Meeting and Exhibit, AIAA 93-0459, Reno/NV, 11–14 January 1993

70. Mankbadi, R.R., Hayder, M.E., Povinelli, L.A.: Structure of supersonic jet flow and its radiated sound. AIAA J. **32**(5), 897–906 (1994)

71. Mankbadi, R.R., Shih, S.-H., Hixon, R., Povinelli, L.A.: Use of linearized Euler equation for supersonic jet noise prediction. AIAA J. **36**(2), 140–147 (1998)

72. Mankbadi, R.R., Shih, S.-H., Hixon, R., Povinelli, L.A.: Direct computation of jet noise produced by large-scale asymmetric structures. J. Propulsion Power **16**(2), 207–215 (2000)

73. Marte, J.E., Kurtz, D.W.: A review of aerodynamic noise from propellers, rotors and lift fans. NASA Tech. Report 32-1462 (1970)

74. Marsden, A.L., Wang, M., Mohammmadi, B.: Shape optimization for aerodynamic noise control. Center for Turbulence Research Annual Brief, pp. 241–47, 2001

75. Mihaescu, M., Szasz, R.-Z., Fuchs, L., Gutmark, E.: Numerical investigations of the acoustics of a coaxial nozzle. 43rd Aerospace Sciences Meeting, 10–13 Jan. 2005, Reno/NV, AIAA 2005-420 (2005)

76. Mitchell, B.E., Lele, S.K., Moin, P.: Direct computation of machwave radiation in an axi-symmetric jet. AIAA J. **35**(10), 1574–1580 (1997)

77. Mitchell, B.E., Lele, S.K., Moin, P.: Direct computation of sound generated by vortex pairing in an axi-symmetric field. J. Fluid Mech. **383**, 113–142 (1999)

78. Mollo-Christensen, E.: Jet noise and shear flow instability seen from experimenter's view point. J. Appl. Mech. Trans. ASME 1 (1967)

79. Moin, P., Squires, K., Cabot, W., Lee, S.: A dynamic sub-grid scale model for compressible turbulence and salar transport. Phys. Fluids A **3**(11), 2746–2757 (1991)

80. Morgans, A.S., Karabasov, S.A., Dowling, A.P., Hynes, T.P.: Transonic helicopter noise. AIAA J. **43**(7), 1512–1524 (2005)

81. Morley, C.I.: How to reduce the noise of jet engines with a method using non-linear disturbance operations. Engineering 783 (1964)

82. Morris, P.J., Long, L.N., Bangalore, A., Wang, Q.: A parallel three dimensional computational aeroacoustics method using non-linear disturbance operations. J. Comput. Phys. **133**, 56–74 (1997)

83. Nagamatsu, H.T., Horvey, G.: Supersonic jet noise. 8th AIAA Aerospace Sciences Meeting, 19–21 Jan., 1970, AIAA Paper 237 (1970)

84. Owis, F., Balakumar, P.: Evaluation of boundary conditions and numerical schemes for jet noise computation. 38th Aerospace Sciences Meeting and Exhibit, Reno/NV, 10–13 Jan. 2000, AIAA Paper 2000-0923 (2000)

85. Ozonik, Y., Long, L.N.: Comparison of sound radiating from engine inlet. AIAA J. **34**(5), 894–901 (1996)

86. Ozonik, Y.: Multi-grid acceleration of a high resolution computational aeroacoustic scheme. AIAA J. **35**(3), 894–901 (1996)

87. Pan, F.L.: The use of surface integral methods in computational jet aeroacoustics. Ph.D. thesis, Purdue University, 2004

88. Pan, F.L., Uzun, A., Lyrintzis, A.S.: Surface integral methods in jet aeroacoustics: refraction corrections. AIAA J. **45**(2), 381–87 (2008)

89. Parthasarathy, S.P., Massier, P.F., Cuffel, R.F.: Comparison of results obtained with various sources used to measure fluctuating quantities. AIAA Aeroacoustics Conference, AIAA Paper 73-1043, Seattle, WA, 15–17 October 1973

90. Powell, A.: Similarity and turbulent jet noise. J. Acoust. Soc. Am. **31**(6), 812–813 (1959)

91. Proudman, I.: The generation of noise by isotropic turbulence. Proc. R. Soc. **214A**, 119–132 (1952)

92. Quemere, P., Sagaut, P.: Zonal multi-grid RANS/LES simulation of turbulent flows. Int. J. Numer. Methods Fluids (2002)
93. Randall, D.G.: Methods for estimating distribution and intensities of sonic bangs. Aeron. Res. Council, London, **R and M 3113** (1957)
94. Rao, P.S.: Supersonic bangs. Aeronaut. Q. **2**, part I: 21, part II: 131 (1956)
95. Ribner, H.S.: Quadrupol correlations governing the pattern of jet noise. J. Fluid Mech. **38**, Part 1, 1–24 (1969)
96. Ribner, H.S.: The noise of aircraft. 4th Congress of International Council of Aeronautical Sciences, Paris, p. 13. Macmillan, London (1969)
97. Roe, P.L.: Technical prospects for computational aeroacoustics. AIAA 92-02-032 (1992)
98. Schumann, U.: Subgrid-scale model for finite difference simulations of turbulent flows in plane channels and annuli. J. Comput. Phys. **86**, 376–404 (1975)
99. Schlinker, R.H., Simonich, J.C., Sharon, D.W., Reba, R.A., Colonius, T., Gudemendsson, K., Ladeincla, F.: Supersonic jet noise from round and chevron nozzles: experimental studies. 30th AIAA Aeroacoustics Conference, AIAA 2009-3257, Miami/FL, 11–13 May 2009
100. Shur, M.L., Spalart, P.R., Strelets, M.K., Travin, A.K.: A hybrid RANS-LES approach with delayed-DES and wall-modelled LES capabilities. Int. J. Heat Fluid Flow **29**, 1638–1649 (2008)
101. Seror, C., Sagaut, P., Bailley, C., Juve, D.: Subgrid-scale contribution to noise prediction and decaying turbulence. AIAA J. **38**(10), 1795–1803 (2000)
102. Seror, C., Sagaut, P., Bailley, C., Juve, D.: On the radiated noise computed in large eddy simulation. Phys. Fluids **13**(2), 476–487 (2001)
103. Smagorinsky, J.: General circulation, experiments with the primitive equations. Mon. Weather Rev., Washington/DC **91**, 99–164 (1963)
104. Spalart, P.R.: Direct simulation of a turbulent boundary layer up to $Re_\theta = 1410$. J. Fluid Mech. **187**, 61–98 (1988)
105. Spalart, P.R., Jou, W.H., Strelets, M., Allmaras, S.R.: Comments on the feasibility of LES and on a hybrid RANS/LES approach. 1st AFOSR International Conference on DNS/LES, Ruston, LA (1997)
106. Spalart, P.R.: Strategies for turbulence modeling and simulation. Int. J. Heat Fluid Flow **21**, 252–263 (2000)
107. Speziale, C.G.: Computing non-equilibrium flows with time-dependent RANS and VLES. 15th International Conference on Numerical Methods in Fluid Mechanics, Monterey, 1996
108. Strawn, R.C., Oliker, L., Biswas, R.: New computational methods for the prediction and analysis of helicopter noise. 2nd AIAA/CEAS Aeroacoustics Conference, AIAA paper 1996-1696, State College, PA, 6–8 May 1996
109. Strelets, M.: Detached eddy simulation of massively separated flows. Aerosciences Meeting, AIAA Paper 2001-0879 (2001), Reno, NV, 3–8 January 2011
110. Tam, C.K.W.: Supersonic jet noise generated by large scale disturbance. AIAA Aeroacoustic Conference, AIAA Paper 73-992, Seattle/WA, 15–17 Oct 1973
111. Tam, C.K.W., Webb, J.C.: Dispersion-relation-preserving finite difference schemes for computational aeroacoustics. J. Comput. Phys. **107**, 262–281 (1993)
112. Taylor, G.I.: Diffusion by continuous movements. Proc. Lond. Math. Soc., Ser. 2 **20**, 196–212 (1921)
113. von Terzi, D.A., Fasel, H.F.: A new flow simulation methodology applied to the turbulent facing step. 40th Aerospace Sciences Meeting and Exhibit, Reno/NV, 14–17 Jan. 2002, AIAA Paper 2002-0429 (2002)
114. Thomas, J.P., Roe, P.L.: Development of non-dissipative numerical schemes for computational aeroacoustics. AIAA 93-3382-CP (1993)
115. Thompson, K.W.: Time-dependent boundary conditions for hyperbolic systems. J. Comput. Phys. **68**, 1–24 (1987)
116. Thompson, K.W.: Time-dependent boundary conditions for hyperbolic systems II. J. Comput. Phys. **89**, 439–461 (1990)

117. Travin, A., Shur, M., Strelets, M., Spalart, P.R.: Detached-eddy simulations past a circular cylinder. Int. J. Flow Turb. Combust. **21**, 252–263 (2000)
118. Troff, B., Manoha, E., Sagaut, P.: Trailing edge noise prediction using LES and acoustic analogy, AIAA J., **38**(4), 575–583 (2000)
119. Uzun, A., Blaisdell, G.A., Lyrintzis, A.S.: Recent progress towards large Eddy simulation code for jet aeroacoustics. 8th AIAA/CEAS Conference and Exhibit, June 17–19, 2002, AIAA Paper 2002-2598 (2002)
120. Uzun, A., Blaisdell, G.A., Lyrintzis, A.S.: 3D large Eddy simulation for jet aeroacoustics. 9th AIAA/CIAS Aeroacoustics Conference and Exhibit, 12–14 May 2003, Hilton Head/SC, AIAA Paper 2003-3322 (2003)
121. Uzun, A.: Ph.D. Thesis, Purdue University, 2003
122. Uzun, A., Blaisdell, G.A., Lyrintzis, A.S.: Sensitivity to the Smagorinsky constant in turbulent jet simulation. AIAA J. **41**(10), 2077–2079 (2003)
123. Uzun, A., Lyrintzis, A.S., Blaisdell, G.A.: Coupling of integral acoustic methods with LES for jet noise prediction. Aeroacoustics **3**(4), 297–346 (2005)
124. Uzun, A., Blaisdell, G.A., Lyrintzis, A.S.: Impact of subgrid-scale models on jet turbulence and noise. AIAA J. **44**(6), 1365–1368 (2006)
125. Uzun, A., Hussain, M.Y.: Noise generation in the near nozzle region of a chevron nozzle jet flow. 28th AIAA Aerospace Conf., May 2007, AIAA Paper 2007-3596 (2007)
126. Wang, M., Moin, P.: Computation of the trailing edge flow and noise using large eddy simulation. AIAA J. **38**(12), 2201–2209 (2000)
127. Warren, C.H.E., Randall, D.G.: The theory of sonic bangs. Prog. Aeron. Sci. **1**, 238–274 (1961)
128. Whitham, G.B.: The flow pattern of a supersonic projectile. Comm. Pure Appl. Math. **5**, 301 (1952)
129. Yoshizawa, A.: Statistical theory for compressible turbulent shear flows. Phys. Fluids **29**(7), 2152–2164 (1986)

List of Symbols

The more commonly used symbols are given below followed by units.

$\mathbf{A}$	A matrix vector
A	Amplitude
a	Heat diffusivity coefficient, m^2s^{-1}
$\mathbf{B}$	A matrix vector
c	Sonic speed, ms^{-1}
c_p	Specific heat at constant pressure, $\text{J}\,\text{kg}^{-1}\text{K}^{-1}$
C_S	Smagorinsky constant
c_μ	A turbulence constant
$c_{\mu 1}, c_{\mu 2}$	Turbulence constants
D	Diameter, m
$\mathbf{E}$	x-directional vector
E	Volumetric energy density, Jm^{-3}
$\mathbf{F}$	y-directional vector
F	Force, N
G	Green's function
g	Gravitational acceleration, ms^{-2}
$\mathbf{H}$	A vector in flow equation
I	Sound intensity, Wm^{-2} or dB
k	Heat conductivity coefficient, $\text{Wm}^{-1}\text{K}^{-1}$
k	Order of frequency
L	Length, m
L_c	Jet core length, m
L_s	Supersonic mixing zone length, m
$\mathbf{L}_s$	Separation vector in a moving coordinate, m
l	Length scale
M	Mach number
$\mathbf{M}_c$	Convective Mach number
m	String mass per unit length, kgm^{-1}
m	Molemass, $\text{kg}\,\text{kmole}^{-1}$

T. Bose, *Aerodynamic Noise: An Introduction for Physicists and Engineers*,
Springer Aerospace Technology 7, DOI 10.1007/978-1-4614-5019-1,
© Springer Science+Business Media New York 2013

Pr	Prandtl number
P	Acoustic power,
p	Pressure, Nm^{-2}
q	Source
$\dot{Q}_m$	Volumetric mass source, $\mathrm{kgm}^{-3}\mathrm{s}^{-1}$
$\dot{q}_m$	Mass source, kgs^{-1}
R	Radius, m
R	Correlation
R	Gas constant, $\mathrm{Jkg}^{-1}\mathrm{K}^{-1}$
$R*$	Universal gas constant, $\mathrm{Jkmole}^{-1}\mathrm{K}^{-1}$
r	Radial distance, m
$\mathbf{S}$	Source function
S	Stress tensor
S	Surface area, m^2
s	Entropy, $\mathrm{J\,kg}^{-1}\,K^{-1}$
T	Temperature, K
T_o	Stagnation temperature, K
T	Tension, N
T	Period, quadrupole moment
T	Quad
T	Time scale, s
t	Time, s
$\mathbf{U}_c$	Convection velocity, ms^{-1}
u	Gas velocity in x coordinate direction, ms^{-1}
$\dot{V}$	Volumetric source strength, $\mathrm{m}^3\mathrm{s}^{-1}$
v	Gas velocity in y coordinate direction, ms^{-1}
W	Probability
w	Gas velocity in z coordinate direction, ms^{-1}
x, y, z	Coordinate directions in Cartesian coordinates, ms^{-1}
$\mathbf{x}$	Point of observation
$\mathbf{y}$	Point of source
$\mathbf{y}$	Point of source displaced
β	Coefficient for loss of sound
Δ	Cut-off length, m
δ	Dirac delta function
δ	Separation vector, m
ε_m	Eddy diffusivity coefficient, $\mathrm{m}^2\mathrm{s}^{-3}$
ε	Turbulent dissipation, s^{-1}
φ	Angle
γ	Isentropic coefficient
Λ	A diagonal vector
η	Kolmogorov length scale
λ	Wavelength, m
μ	Dynamic viscosity coefficient, $\mathrm{kgm}^{-1}\mathrm{s}^{-1}$
μ_t	Turbulent viscosity coefficient, $\mathrm{kgm}^{-1}\mathrm{s}^{-1}$

ν	Frequency, s^{-1}
ν	Kinematic viscosity coefficient, m^2s^{-1}
Ω	Volume, m^3
ω	Radian frequency, rad.s^{-1}
ρ	Density, kgm^{-3}
σ	Favre-filtered viscous stress tensor
σ_k	Prandtl number of turbulent kinetic energy
σ_ε	Prandtl number of turbulent dissipation
θ	Angle
τ	Shear stress, Nm^{-2}
τ^*	Shear stress, Nm^{-2}
τ	Viscous stress tensor, Nm^{-2}
τ	Time delay, s
$<>$	Average

Superscripts

$(\,)'$	Perturbed quantity
$(\,)''$	Perturbed quantity from mass-averaged quantity
$\overline{(\,)}$	Average or mean variable
$\widetilde{(\,)}$	Mass-averaged variable

Subscripts

∞	Free-steam condition
o	Gas at rest

Index

A

A-weighting method, 22
absorbing frequency, 146
acoustic boundary layer thickness, 12
acoustic lining, 146
acoustic power, 21, 63, 79, 134
acoustic power of a longitudinal quadrupole,
    50
acoustic power of a single dipole, 43
acoustic power of the fluctuating longitudinal
    quadrupole, 48
acoustic power spectrum, 63
acoustic wave motion, 19
active noise control, 145
advance ratio, 123
aerodynamic load vector, 122
aeroelastic problems, 121
aircraft wing trailing edge, 84
airframe noise, 142
airport noise, 141
anechoic chamber, 13, 149
angular directivity pattern, 69
approaching flow Mach numbers, 123
artificial velocity fluctuations, 117
attached eddies, 111
autocorrelation, 19, 65

B

basic governing equations, 89
beat instruments, 26
beats, 7
Bessel equation, 106
Bessel function of first order, 40
blade-cutting noise, 146
bluff-body noise, 142
Bose line of headphones, 145

boundary layer thickness, 13
boundary layer type flow, 92
boundary-layer approximation, 90
broadband frequency spectrum, 84
broadband noise, 21, 129, 146
buzz-saw noise, 146

C

calibration, 140
cancellation effects in dipole, 43
chevron nozzle, 143
choking, 25
chord angles, 123
chromatic scale, 27
coefficient for loss of sound, 12
coefficient for the loss of sound, 12
combustion noise, 83
complex eigenvalue, 106
computational aeroacoustics, 115
cone of silence, 82, 99
control surface, 138
conventional Smagorinsky model, 109
correlation integral, 66
correlation volume, 18, 59, 78
creep and fatigue failures, 121
crossed laser beams, 16
Curle equation, 95
cutoff Mach number, 139
cylindrical Rayleigh equation, 106

D

decay time of eddies, 81
delayed detached eddy simulation (DDES),
    112
delta function, 7, 59, 108

T. Bose, *Aerodynamic Noise: An Introduction for Physicists and Engineers*,
Springer Aerospace Technology 7, DOI 10.1007/978-1-4614-5019-1,
© Springer Science+Business Media New York 2013

MIX
Papier aus verantwortungsvollen Quellen
Paper from responsible sources
FSC® C105338

If you have any concerns about our products,
you can contact us on
ProductSafety@springernature.com

In case Publisher is established outside the EU,
the EU authorized representative is:
Springer Nature Customer Service Center GmbH
Europaplatz 3, 69115 Heidelberg, Germany

Printed by Libri Plureos GmbH
in Hamburg, Germany